Digital Technology and Sustainability

This book brings together diverse voices from across the field of sustainable human-computer interaction (SHCI) to discuss what it means for digital technology to support sustainability and how humans and technology can work together optimally for a more sustainable future.

Contemporary digital technologies are hailed by tech companies, governments and academics as leading-edge solutions to the challenges of environmental sustainability; smarter homes, more persuasive technologies, and a robust Internet of Things hold the promise for creating a greener world. Yet, deployments of interactive technologies for such purposes often lead to a paradox: they algorithmically "optimize" heating and lighting of houses without regard to the dynamics of daily life in the home; they can collect and display data that allow us to reflect on energy and emissions, yet the same information can cause us to raise our expectations for comfort and convenience; they might allow us to share best practice for sustainable living through social networking and online communities, yet these same systems further our participation in consumerism and contribute to an ever-greater volume of electronic waste. By acknowledging these paradoxes, this book represents a significant critical inquiry into digital technology's longer-term impact on ideals of sustainability.

Written by an interdisciplinary team of contributors, this book will be of great interest to students and scholars of human-computer interaction and environmental studies.

Mike Hazas is Senior Lecturer in the School of Computing and Communications, Lancaster University.

Lisa P. Nathan is Associate Professor at the School of Library Archival and Information Studies, University of British Columbia.

Routledge Studies in Sustainability

www.routledge.com/Routledge-Studies-in-Sustainability/book-series/RSSTY

Human Rights and Sustainability
Moral responsibilities for the future
Edited by Gerhard Bos and Marcus Düwell

Sustainability and the Art of Long Term Thinking
Edited by Bernd Klauer, Reiner Manstetten, Thomas Petersen and Johannes Schiller

Cities and Sustainability
A new approach
Daniel Hoornweg

Transdisciplinary Research and Practice for Sustainability Outcomes
Edited by Dena Fam, Jane Palmer, Chris Riedy and Cynthia Mitchell

A Political Economy of Attention, Mindfulness and Consumerism
Reclaiming the Mindful Commons
Peter Doran

Sustainable Communities and Green Lifestyles
Consumption and Environmentalism
Tendai Chitewere

Aesthetic Sustainability
Product Design and Sustainable Usage
Kristine H. Harper

Stress, Affluence and Sustainable Consumption
Cecilia Solér

Digital Technology and Sustainability
Engaging the Paradox
Edited by Mike Hazas and Lisa P. Nathan

Digital Technology and Sustainability

Engaging the Paradox

Edited by
Mike Hazas and Lisa P. Nathan

LONDON AND NEW YORK

First published 2018 by Routledge

2 Park Square, Milton Park, Abingdon, Oxfordshire OX14 4RN
52 Vanderbilt Avenue, New York, NY 10017

Routledge is an imprint of the Taylor & Francis Group, an informa business

First issued in paperback 2019

Copyright © 2018 selection and editorial matter, Mike Hazas and Lisa P. Nathan; individual chapters, the contributors

The right of Mike Hazas and Lisa P. Nathan to be identified as the authors of the editorial material, and of the authors for their individual chapters, has been asserted in accordance with sections 77 and 78 of the Copyright, Designs and Patents Act 1988.

All rights reserved. No part of this book may be reprinted or reproduced or utilised in any form or by any electronic, mechanical, or other means, now known or hereafter invented, including photocopying and recording, or in any information storage or retrieval system, without permission in writing from the publishers.

Notice:
Product or corporate names may be trademarks or registered trademarks, and are used only for identification and explanation without intent to infringe.

British Library Cataloguing-in-Publication Data
A catalogue record for this book is available from the British Library

Library of Congress Cataloging-in-Publication Data
A catalog record for this book has been requested

ISBN: 978-1-138-20588-8 (hbk)
ISBN: 978-0-367-27116-9 (pbk)

Typeset in Times New Roman
by Apex CoVantage, LLC

To current and future SHCI scholars and practitioners,
and our children

Contents

List of table	x
List of figures	xi
Acknowledgements	xii
List of contributors	xiii

Photo essay 1: Selfie time   1
ELI BLEVIS

Introduction: Digital technology and sustainability: engaging the paradox   3
MIKE HAZAS AND LISA P. NATHAN

Photo essay 2: Artifice and nature   14
ELI BLEVIS

PART 1
Assessing the field   15

1 **Three principles of sustainable interaction design, revisited**   17
DAVID ROEDL, WILLIAM ODOM, AND ELI BLEVIS

2 **Towards a social practice theory perspective on sustainable HCI research and design**   31
ADRIAN K. CLEAR AND ROB COMBER

3 **A conversation between two sustainable HCI researchers: the role of HCI in a positive socio-ecological transformation**   44
SAMUEL MANN AND OLIVER BATES

Response 1a: Sustainable HCI: from individual to system   61
CHRIS PREIST

Response 1b: Sustainability within HCI within society: improvisations, interconnections and imaginations   63
JANINE MORLEY

Photo essay 3: Rooftop garden   67
ELI BLEVIS

PART 2
Addressing limits   69

4   Every little bit makes little difference: the paradox within SHCI   71
SOMYA JOSHI AND TERESA CERRATTO PARGMAN

5   Developing a political economy perspective for sustainable HCI   86
BONNIE NARDI AND HAMID EKBIA

6   Software engineering for sustainability: tools for sustainability analysis   103
BIRGIT PENZENSTADLER AND COLIN C. VENTERS

Response 2: Challenging the scope?   122
ENRICO COSTANZA

Photo essay 4: Classroom exercise   125
ELI BLEVIS

PART 3
Ways to engage with others   127

7   Communicating SHCI research to practitioners and stakeholders   129
CHRISTIAN REMY AND ELAINE M. HUANG

8   Negotiating and engaging with environmental public policy at different scales   140
VANESSA THOMAS

9   On the inherent contradictions of teaching sustainability at a technical university   154
ELINA ERIKSSON AND DANIEL PARGMAN

10 Participation in design for sustainability   166
    JANET DAVIS AND SANDRA BURRI GRAM-HANSEN

    Response 3a: Connected and complicit   179
    MÉL HOGAN

    Response 3b: From participatory design to participatory
    governance through sustainable HCI   182
    RÓNÁN KENNEDY

    Photo essay 5: Airstream   185
    ELI BLEVIS

PART 4
Inspiring futures   187

11 A sustainable place: everyday designers as placemakers   189
    AUDREY DESJARDINS, XIAOLAN WANG, AND RON WAKKARY

12 Interaction design for sustainability futures: towards
    worldmaking interactions   205
    ROY BENDOR

13 Think local act local: the case of Burning Man   217
    A.M. TSAASAN AND BONNIE NARDI

    Response 4: Sustainability futures and the future of
    sustainable HCI   231
    YOLANDE STRENGERS

    Photo essay 6: Locked gate   234
    ELI BLEVIS

    Epilogue   235
    MIKE HAZAS AND LISA P. NATHAN

    Index   237

Table

6.1 Stakeholders affected by sustainability dimensions 112

Figures

3.1	One CHI-soaked evening	45
3.2	Mann-Bates sustainable HCI rubric	57
6.1	Requirements engineering for sustainability (with highlighted artifacts, stakeholder model, and sustainability analysis for the focus of this chapter)	109
6.2	Ways for stakeholder identification	110
6.3	Stakeholder model of the resilient smart garden system	114
6.4	Sustainability analysis of the resilient smart garden system	115
11.1	Kitchen unit evolution	196
11.2	A gardener cleaning communal place in his garden	197
11.3	(Left) Storage and sleeping platform (Right) Benches and table built on top of the storage platform	198
11.4	Rebuilt and decorated fence in the community garden	199
11.5	(Left) Hand-installed hooks to the ceiling (Right) 3D model of the benches and table on the platform	200
11.6	(Left) Communal plot of a community garden (Right) Harvest-sharing shelf	201
13.1	Mini-effigy burn and creation	218
13.2	A crowd gathered for the Burning Man	219
13.3	A participant learning to adjust the torch while practicing welds on a thick piece of steel that is resting on one of the former food storage containers	226

Acknowledgements

We thank all past and present contributors to the SIGCHI HCI & Sustainability Community, particularly those who have organized or attended special interest group meetings, panels, and workshops.

Contributors

Editors

Mike Hazas (Senior Lecturer, School of Computing and Communications, Lancaster University). Mike works at the interface of human-computer interaction and studies of social practice, striving for nuanced understandings of sustainability through both qualitative and quantitative data. Inspiration: Bob the Minion

Lisa P. Nathan (Assistant Professor, School of Library Archival and Information Studies, University of British Columbia). Lisa is privileged to collaborate with fantastic people to (re)imagine and (re)design information practices – ways of stewarding information – to address long-term challenges (e.g., decolonization, social justice, environmental resilience). Inspiration: Kim Lawson

Chapter authors

Oliver Bates (Senior Research Associate, School of Computing and Communications, Lancaster University). Oliver's research looks to understand how digital technology is situated in practices, how practices impact sustainability, and how we can redesign digital technology for more sustainable trajectories. Inspiration: Billy Bragg

Roy Bendor (Assistant Professor, Department of Industrial Design, Delft University of Technology). Roy explores the design and use of interactive media as means to disclose, provoke, and reshape our political imagination. Inspiration: John Robinson

Eli Blevis (Professor of Informatics, School of Informatics and Computing, Indiana University Bloomington; Visiting (ongoing) and Adjunct (appointed) Professor of Interaction Design, School of Design, The Hong Kong Polytechnic University). Eli is best known for his work on sustainable interaction design. He also engages visual thinking – especially photographic foundations of HCI, and design theory – especially transdisciplinary design. Inspiration: Shunying An Blevis

xiv *Contributors*

Teresa Cerratto Pargman (Associate Professor, Department of Computer and Systems Sciences, Stockholm University). Teresa works within the field of human-computer interaction, exploring issues of civic participation and sustainability in a future of economic and ecological limits. Inspiration: my children

Adrian K. Clear (Senior Research Fellow in Digital Living, Department of Computer and Information Sciences, Northumbria University). Adrian's work takes a social practice orientation to understanding how technology impacts everyday life, and how design might support and promote more sustainable ways of living. Inspiration: John Dewey

Rob Comber (Lecturer, School of Computing Science, Newcastle University). Rob's human-computer interaction research uses a lens of 'designing for community' to examine topics such as activism, citizen science, community education, and food and technology.

Janet Davis (Associate Professor, Department of Mathematics and Computer Science, Whitman College). Janet pioneered the application of participatory design to persuasive technology. Now she is contemplating the moral risks and opportunities of using information technology to influence how people use language. Inspiration: Joseph M. Williams

Audrey Desjardins (Assistant Professor, School of Art + Art History + Design, University of Washington). Audrey uses design as a way to critically reflect on people's creative tactics to make home and to investigate potential futures in domestic spaces. Inspiration: Traveling through multiple US national parks

Hamid Ekbia (Professor, School of Informatics and Computing, Indiana University, Bloomington). Hamid is interested in the political economy of computing and in how technologies mediate cultural, socio-economic, and geo-political relations of modern societies. Inspiration: Karl Marx

Elina Eriksson (Assistant Professor, School of Computer Science and Communication, KTH Royal Institute of Technology). Elina is interested in the social construction of denial and wants more people to engage in sustainability issues – even if it hurts. Inspiration: Kari Marie Norgaard

Sandra Burri Gram-Hansen (Assistant Professor, Department of Communication and Psychology, Aalborg University, Denmark). Sandra's work focuses on persuasive design, co-creation, participatory design processes, and applied ethics. She is particularly interested in climate communication and sustainability, as well as designs that bridge between physical and digital realms. Inspiration: J.K. Rowling

Elaine M. Huang (Associate Professor, Department of Informatics, University of Zurich). Elaine leads the People and Computing group and conducts HCI and ubiquitous computing research that is driven by deep inquiry into human practices and communication. Inspiration: Her PhD students

Somya Joshi (Assistant Professor, Department of Computer & Systems Science, Stockholm University). Somya participates in diverse forms of activism and disruption that challenge the current worldview of unlimited resources and incumbent consumption/waste practices. Inspiration: Naomi Klein/Leonard Cohen

Samuel Mann (Professor, CapableNZ, Otago Polytechnic). Sam's focus is making a positive difference through professional practice. He developed the role of the sustainable practitioner, the Sustainable Lens and Transformation Mindset. Inspiration: People making a difference, some of whom I've collaborated with on SustainableLens.org

Bonnie Nardi (Professor, Department of Informatics, University of California, Irvine). Bonnie wants to know how we can design a future that lets us keep the Internet, take from the rich and give to the poor, and have a lot of fun. Inspiration: André Gorz

William Odom (Assistant Professor, School of Interactive Arts and Technology, Simon Fraser University). William investigates how (if) we can create long-term strategies for sustaining viable human and environmental futures. Inspiration: Tony Fry

Daniel Pargman (Associate Professor, School of Computer Science and Communication, KTH Royal Institute of Technology). Daniel strives to understand how the future clash between the exponential growth of computing and the limited resources of a finite planet will unfold. Inspiration: the de-growth movement

Birgit Penzenstadler (Assistant Professor, Department of Computer Engineering and Computer Science, California State University Long Beach). Birgit works on using requirements engineering techniques for developing software systems for sustainability. She is also interested in quality assurance, innovation, and open source development. Inspiration: Sara Bareilles' song "Brave"

Christian Remy (Postdoctoral researcher, Department of Informatics, University of Zurich). Christian recently investigated issues of sustainability by trying to tackle e-waste and obsolescence through sustainable interaction design. Inspiration: hiking in the mountains

David Roedl (Lead UX Designer, 3M Design). David studies relationships among technology, culture, political economy, and the environment; he designs digital products and services with hope for a just and sustainable future. Inspiration: David Hakken and David Harvey

Vanessa Thomas (PhD Candidate, HighWire Centre for Doctoral Training, Lancaster University). Vanessa researches social practices, public policies, and the environmental footprint of digital technologies. She also co-runs the Lickable Cities project and loves cake. Inspiration: Sara Ahmed

a.m. tsaasan (PhD Student, Department of Informatics, University of California, Irvine). Marie studies playful learning contexts in local and virtual communities that improve public spaces by increasing inclusion, fun, and personal responsibility. Inspiration: Mariajose Nunes, teacher of islander aesthetics wherein all members are essential

Colin C. Venters (Senior Lecturer in Software Systems Engineering, School of Computing and Engineering, University of Huddersfield). Colin is a founding member of the Karlskrona Consortium for Sustainability Design. His current research focuses on sustainable software systems engineering for computational science and engineering in extreme-scale computing environments from a software architecture perspective. Inspiration: Bonnie "Prince" Billy

Ron Wakkary (Professor, School of Interactive Arts and Technology, Simon Fraser University and Visiting Professor in Industrial Design, Eindhoven University of Technology). Ron's research aims to be reflective and generate design inquiries into technologies and everyday practices. Inspiration: Jane Jacobs

Xiaolan Wang (PhD Candidate, School of Interactive Arts and Technology, Simon Fraser University). Xiaolan studies social innovation projects by applying the theories of infrastructuring, aiming for exploring the potentialities design can have in supporting social change toward sustainability. Inspiration: Christopher Le Dantec

Respondents and advisory board

Enrico Costanza (Lecturer, UCL Interaction Centre, University College London). Currently Enrico's research focuses on designing systems to help people make sense of data, and on interaction with smart and autonomous systems in everyday situations. Inspiration: Bruno Munari

Lorenz M. Hilty (Professor and Sustainability Delegate, University of Zurich; Head of Group at Empa, Swiss Federal Laboratories for Materials Science and Technology). Lorenz is a computer scientist who is continuously looking for ways to reduce the excessive material and energy throughput of our lifestyles. Inspiration: Hans-Christoph Binswanger.

Mél Hogan (Assistant Professor of Environmental Media, Communication, Media and Film, University of Calgary). Mél's work looks at the social implications and environmental impacts of data centers globally. Inspiration: animals

Rónán Kennedy (Lecturer, School of Law, National University of Ireland Galway). Rónán explores how law and information and communications technology influence each other, trying to see through the hype and understand the real consequences. His research focuses on the use of ICT for environmental regulation. Inspiration: Low

Jennifer Mankoff (Associate Professor, Human Computer Interaction Institute, Carnegie Mellon University). Jen's research embodies a human-centered perspective on data-driven applications. Her goal is to combine empirical methods with technological innovation to construct middleware (tools and processes).

Janine Morley (Senior Research Associate, DEMAND Centre, Lancaster University). Janine is a sociologist who works collaboratively to study the relationships between social practices, technology, and sustainability. She is particularly fascinated by digital technologies, social change, and futures. Inspiration: Apus apus

Chris Preist (Reader, Department of Computer Science, University of Bristol). Chris studies the ways in which digital technology changes human behavior and practices, and the effects of this (both positive and negative) on the many challenges of sustainability. Inspiration: Eihei Dogen

Yolande Strengers (Senior Research Fellow, Centre for Urban Research, RMIT University). Yolande is a sociologist of technology and design and co-leader of the Beyond Behaviour Change research program at RMIT. She researches the changing role of digital and smart technologies in everyday life and their implications for energy and sustainability outcomes. Inspiration: her compost bin

Photo Essay 1

Selfie Time (2017). A monolithic LED display advertisement for Samsung proclaims, *"Selfie Time"*. Interaction design is implicated in this advertising use of digital energy infrastructure to promote consumption of ever more digital energy infrastructure and consumer products. Reflection on Bendor: *Interaction design and sustainability futures: Towards worldmaking interactions* (apropos of persuasive interactions); and Thomas: *Negotiating and engaging with public policy at different scales* (apropos of waste electronic equipment); and Joshi: *Every little bit makes little difference: The paradox within SHCI* (apropos of discourse of consumption and technological innovation); and Roedl, Odom, and Blevis: *Three Principles of Sustainable Interaction Design, Revisited* (apropos of invention and disposal and apropos of references to emerging work on digital energy infrastructure).

Eli Blevis

Introduction
Digital technology and sustainability: engaging the paradox

Mike Hazas and Lisa P. Nathan

Sustainable human-computer interaction: a field of contradictions

In an era of high-speed Internet, ever-smarter devices, and incessant social media updates, rapid change becomes the norm, at least for individuals with access to these technologies. Yet, certain transformations are happening more quickly – and at a larger scale – than can be normalized. Ancient ice sheets are melting and oceans are rising. Storm systems around the world are growing larger, stronger, and more destructive. In response, tech companies and governments advocate for digital technological innovation as a solution to climate degradation; smarter homes, persuasive technologies, and a robust Internet of Things promise a more sustainable future for humanity. However, deployments of "sustainable" interactive technologies often reveal a paradox: they collect and display data that allow us to reflect on energy and emissions, yet the same information can raise our expectations for comfort and convenience; they algorithmically "optimize" heating and lighting for idealized houses with little regard to the non-optimized fluctuations across the diversity of life in our homes; they might enable us to develop practices for sustainable living through social networking and online communities, yet the perpetual versioning of these systems demands our participation in consumerism and generates an inconceivable amount of toxic waste.[1]

This text brings together diverse voices from across the nascent field of sustainable human-computer interaction (SHCI): scholars, researchers, and practitioners who study, envision, create, and critique interactive digital technologies that engage the issue of sustainability. Through their contributions, chapter authors acknowledge and address both the promise and the paradox of addressing sustainability concerns through innovation in digital technology. The question of sustainability, what people around the world want to sustain – has diverse – and often contradictory responses. We posit that every sustainability discussion is tied to the question "What is being valued?" – even in conversations that support climate change denial. Through complex interactions around the world, possibilities for human life (and that of other life forms) on this planet constantly shift, expanding, and contracting. As editors of this collection, we did not seek a singular, technologically deterministic "answer" to the myriad problems related to climate change. Rather, our primary objective in developing this collection was to create space for critical inquiry into digital technology's longer-term influence on the quality of our lives, and the lives of future generations.

Informed by a process of sustained engagement among the book's contributors and other experts in the field, chapter authors address fundamental questions: What does it mean for digital technologies to support sustainability? What are the appropriate, meaningful metrics to use when evaluating digital systems for sustainability? What changes are necessary for our entanglements with digital technologies to lead towards a reduction, rather than growth, in greenhouse gas emissions, energy demand, and e-waste? What new possibilities for digital technologies are wholly unrealized, and what futures have yet to be envisioned? In a world undergoing dramatic ecological transformations, how might our use of digital technologies in their many forms lessen suffering and enhance humanity's capacity to thrive?

Cultivating deliberation and reflection

As editors we wanted to bring forward innovative approaches and bring together scholars working in this area to question and strengthen each other's work. We were not interested in an edited book of stand-alone, siloed contributions. Instead, we designed a process that facilitated ongoing interaction, constructive discussion and iteration.

Early in our planning, we (the editors) formed an Advisory Board of scholars well positioned to provide feedback and potentially critique chapters or sections of the book. We posted an open call for chapter contributors to create new writing on or related to SHCI. We invited authors whose perspectives could be in discussion with each other, highlighting different viewpoints and approaches to this area of enquiry. After initial full drafts of accepted chapters were submitted, each was assigned at least two reviewers, including Advisory Board members and other contributors. In addition to written feedback, reviewers were asked to meet virtually or in person with chapter authors to share reactions to the full drafts and discuss ideas for strengthening the work. Towards the end of the chapter-reviewing process, we also hosted four virtual discussion groups, inviting all contributors to share insights from their chapters and to debate questions related to the book in its entirety.

Once authors submitted their final chapters, we asked members of the Advisory Board and a few sustainability scholars, to respond to one of the four sections of the book. Once these responders contributed their pieces, chapter authors were provided an opportunity to comment on these responses. This process invited fresh perspectives from specialists reading across chapters and helped us understand the chapters' contributions from different positions. This iterative process of feedback between authors, Advisory Board, editors and respondents demonstrated how paradoxes (and conflict) are part and parcel of sustainability discussions.

Through this introductory chapter we offer a brief contextualizing history, and review of each section of the book. Yet, before guiding you through the four sections, we draw your attention to the six photo essays that invite contemplation on themes found throughout the chapters. These photo commentaries were generously created and shared by Eli Blevis. In addition to being an avid photographer, Blevis is also a design researcher and author of the most highly cited SHCI article to date (Blevis, 2007). Through his images and their accompanying analyses, Blevis connects scenes of contemporary life in affluent societies with uncomfortable truths discussed in chapters across this collection. His photo commentaries assist in making scholarly insights more accessible and mundane scenes of daily life generative. We encourage you to revisit and reconsider these commentaries as you make your own way through the chapters and the responses.

Key reflections on the past 10 years of sustainable HCI

Within the broader field of HCI, engagement with issues related to sustainability significantly increased in momentum and volume after three events at the SIGCHI Conference on Human Factors in Computing Systems (CHI) in 2007: (1) the presentation by Eli Blevis of his seminal SHCI paper (Blevis, 2007); (2) the inaugural gathering of the ACM SIGCHI HCI & Sustainability Community;[2] and (3) the establishment of the mailing list sustainable-chi@googlegroups.com (in 2017 over 500 members). Since CHI 2007, there has been increasing activity in the area, evidenced by more published papers, workshops, new conference series, and journal special issues. In the remainder of our introduction, we provide the reader with hints of the varied perspectives within the collection, positioning chapters alongside prior debates and events in sustainable HCI. This is to give context to the field of HCI, and this particular text, expressly for readers unfamiliar with the field and its brief history.

As mentioned above, many SHCI papers reference Blevis (2007) as inspiration; it is a provocative piece that grappled with how HCI and design might take into account concerns of sustainability. Since it was published, few have sought to critically evaluate, much less update, the "sustainable interaction design" framework originally proposed. In Chapter 1, Blevis returns with lead collaborators Roedl and Odom to give a frank assessment of what has taken place in the literature, with specific regard to three out of the five principles appearing in the 2007 paper: linking invention and disposal; promoting renewal and reuse; and promoting quality and equality. Roedl et al. (this volume) point to more recent work by Preist et al. (2016) that advances the original analysis in Blevis (2007) to hone these three principles beyond digital artifice to the increasingly energy-intensive domain of digital infrastructure.

As sustainable HCI has grown, it has become a confluence of researchers from different research and academic backgrounds: most prominent perhaps are computer science, interaction design, psychology, anthropology, and sociology. One of the earlier approaches identified originates from sociology: to observe and explore social practices more broadly, grounded in data gathered using qualitative methods (Pierce et al., 2010; Strengers, 2011). In Chapter 2, Clear and Comber provide a succinct history of the mobilization of practice approaches within sustainable HCI, and recount the criticisms of interaction design for focusing solely on the individual. The first part of their chapter is helpful for its careful selection of the references that speak most directly to practice and sustainability. Having recounted the case, Clear and Comber then detail the particular challenges for using broader practice framings for intervention design; these include both the practical (how can one designer possibly take all aspects of sustainability into account) as well as the theoretically informed (designers should focus on the linkage between the elements of practice, and how those links change).

As the dialogue on sustainable HCI has developed at conferences and workshops (e.g., Blevis et al., 2015; Huh et al., 2010; Knowles et al., 2016; Silberman et al., 2014a), a palpable sense of frustration, and at times despair, has also grown. In some cases this has inspired attendees to write about their dissatisfaction (e.g., Silberman et al., 2014b) and to pursue ways of collecting, integrating and debating work beyond workshop events (Knowles & Håkansson, 2016). Mann and Bates (Chapter 3) offer a dialogue between two SCHI scholars, literally giving voice (through links to dozens of podcasts) to many of the issues that have come up over the years. As editors, we appreciate the experiment that Mann and Bates make by rejecting the conventional academic chapter format, and we value their contribution particularly for its exposition of the fact that doing sustainability research in HCI (as well as many other areas) can be a vexing business, difficult for a researcher's or practitioner's perseverance and morale.

Preist and Morley, through their respective responses to Part 1, each conclude by commenting on certain features and turns within sustainable HCI; and similarly each provide examples of both new and continued links with other disciplines (e.g., design futures, management science, sociology of energy, economics) as sustainable HCI continues to reflect upon, question, and rework itself.

Grappling with limits

While there is broad acknowledgement that the scope of SHCI investigations should not be limited to isolated interactions and interventions, there is also concern that too many people lack an awareness of the limitations of regional and global resources, and systems of provision. This has been openly discussed since the earliest days of sustainable HCI, and indeed a point of focus by Tomlinson et al. (2012) and the annual Workshop on Computing within Limits (beginning in 2015).[3] Taking "limits" as a point of departure is an important critical perspective for HCI scholars concerned with issues of sustainability. In contrast to technocentric approaches, this sort of lens helps us avoid contributing to futures which are ever more resource-intensive. Joshi and Cerratto Pargman (Chapter 4) concretely connect the approach of recognizing resource and planetary boundaries, to the 10-year history of writing in SHCI. They also identify links to a longer history of thinking about limits in other disciplines, such as economics and political ecology. They utilize an analysis of online data and fieldwork with Fairphone stakeholders to illustrate the central conundrum in sustainable HCI: current notions of improvement and advancement in digital interactive system design on one hand, and limitations to resources, waste, and carbon emissions on the other.

Nardi and Ekbia (Chapter 5) then focus on this systemic issue and consider what we can gain in directly acknowledging that we work within current configurations of political economy that inherently value and foster continual growth, with few meaningful commitments to an aggregate reduction of resource consumption, or to minimizing environmental impact. The sustainable HCI community, according to Nardi and Ekbia, needs to pick the big problems, see how those problems fit into the current political economy, use simple but scientifically-grounded arguments to persuade and educate, and work up bold designs and new practices which are important for developing the future, even if they are not realistic or compatible within current systems. They utilize three case studies and a large number of examples and references to make their proposed avenues for progress clear.

A crucial component of designing within limits is that of software – something too often neglected in HCI's focus on human-machine interactions. Software is implicated in nearly all aspects of daily practice and social processes in industrialized economies (Kitchin & Dodge, 2011). Software runs not only in the visible places like phones, laptops, businesses, and homes, but also behind the scenes in public transportation, automobiles, shipping, stock markets, on-demand video streaming, and the profligate distributed systems (data centers) that prop up the many seemingly free digital services that we rely upon in everyday practice (e.g., self-tracking, navigating, texting, and tweeting). These are typically designed with highly localized digital resources in mind (phone battery life, or data center processing time) but lack a wider view and thus contribute to the continual escalation of Internet-related energy demand and emissions (Hazas, 2016). Penzenstadler and Venters (Chapter 6) discuss what it would mean to engineer, from the early requirements onwards, software which is more regionally and globally sustainable. The key to the goal of incorporating sustainability into software engineering is knowledge from other disciplines (such as lifecycle assessment) and accounting for a much broader set of stakeholders.

In his response, Costanza emphasizes the marked difference in the scales of abstraction and boundaries of concern across the three chapters included in Part 2. There is a necessity of keeping research somewhat partial, in order for it to remain tractable. He wonders if these "limits" of sustainable HCI itself, then, should be acknowledged more explicitly and perhaps embraced.

Beyond HCI audiences

While the chapter authors in Part 2 demonstrate how SHCI scholars are determined to recognize and work *within* planetary resource limits, contributors to Part 3 argue for SHCI researchers to go *beyond* the limits of their accustomed research community when communicating their findings. The academic silo is not an issue that SHCI faces alone. Scientists in the United States recently held a "March for Science" because they fear that their life's work isn't valued outside the academy.

This suggests a troubling misalignment between the beliefs of those committed to the research enterprise and the public's perception of the research establishment. Although other academic disciplines face the problem of communicating with broader audiences, we (the editors) believe that those in the HCI community are a particularly privileged group of scholars. Participating in the community comes with high financial costs, particularly for those who do not live in Canada, the US, the UK, or Western Europe. Simply attending the Association of Computing Machinery (ACM) Conference on Human Factors in Computing Systems is prohibitively expensive; registration alone is over 1,000 US dollars. Conference cost is an easy factor to point out, but there are others that prevent a more diverse group of people from participating in the academic community. Lack of diversity significantly constrains the knowledge and life experiences those within the SHCI community are able to recognize and appreciate. With a few notable exceptions (e.g., Bidwell et al., 2013; Busse et al., 2012) it is difficult to find perspectives in the SHCI canon that are not framed by issues of an affluent, "fat world" context (Olopade, 2014). Recognizing this privilege is relatively easy. The difficult work comes in maintaining the humility to remember the vast differences in conditions and cultures of human life around the globe. The chapters in this section continue the work of broadening SHCI scholarship to have significance beyond the conference hall.

Communicating to practitioners and other stakeholders who are not immersed in the language and culture of HCI generally, and SHCI specifically, is the central concern of Remy and Huang's contribution (Chapter 7). They unpack some of the difficulties one faces when trying to determine whether SHCI research has had any "real-world" influence. They use the concept of knowledge translation to convey the importance of understanding one's desired audience well to better express your findings to them, in terms that they understand and appreciate. Remy and Huang also offer reflections on the theory-practice gap in HCI, the challenges SHCI writers have in defining a specific audience and tangible practices to assist you in sharing your work with others in meaningful ways.

Zooming out on the national and global scales, Thomas (Chapter 8) gives dedicated treatment to working with policy audiences. Drawing upon a healthy set of earlier HCI work on public policy, the chapter points out how sustainable HCI can directly inform national and globally regional polices, particularly on electrical equipment procurement and waste. Thomas gives the detailed status of these policies in different countries and regions, and reflects on what HCI can learn from them, and vice versa. There are practical and professional challenges in engaging with policy, and Thomas concludes with a frank discussion of these, pointing out that policy is an important route to change at scale, something that so often eludes us.

Those who successfully navigate the educational system well enough to be admitted to a technical university are those fortunate few for whom working hard and perseverance appear to have paid off. They may be excellent at writing papers and taking exams, but are they well positioned to look critically at how their professional goals will make them deeply complicit in ecological collapse? Will they be excited to learn about and discuss how the lives of other people (and life forms) are in jeopardy, in large part because of the technologically dependent lifestyles of others? Eriksson and Pargman (Chapter 9) share their experiences trying to engage frankly and constructively with students about the facts of ecological change. Many undergrads are at a time of their lives where they are first experiencing empowerment and independence. Is this the right time to also come to terms with the dire facts of climate disruptions? Eriksson and Pargman propose generative ways of thinking about how university prepares technology-adept, climate-conscious global citizens. They provide compelling examples and reflections that will support others who wish to persevere with similar initiatives at institutions of grounded learning.

As the first chapters in Part 3 explain, communicating meaningfully with a diverse range of stakeholders demands that you do more than merely avoid academic jargon. Even if a broader public understands your work, will they find it significant? Davis and Gram-Hansen (Chapter 10) take on the challenge of relevance as they face the elephant in the lab (or studio): is your work meaningful to those you are 'doing it for'? This chapter provides a motivating case by grounding research in the concerns and values of those whom Davis and Gram-Hansen hope will utilize the project's outputs. Their example combines persuasive technology practices with participatory design methods. Rather than wait until the work is complete to figure out a way to translate it meaningfully, Davis and Gram-Hansen develop an argument to include participants, those you want to benefit from your work, much earlier in the design process.

In her response to the chapters in Part 3, Hogan expresses her familiarity with the particular difficulties of communicating SHCI concerns to undergrads. She also reflects on the enigma each of the chapters puzzles through the disconnect between concerns over the urgency of climate change and the focus of HCI, trying to improve the engagement between humans and technology. In ways that resonate with earlier chapters, she reminds of us of our own responsibilities to contribute to significant and meaningful change, perhaps including the way we think about the field of SHCI.

Kennedy frames his response to Part 3 using terms that are familiar to those who read all four chapters: conflict, control, and communication. Informed by his ideals of a robust democracy, Kennedy creates connections between the chapters that demonstrate a need to make more room for reflection and debate over the values and ethics that underlie the research, teaching, and service of academia.

Alternative futures and design

For design-oriented researchers and practitioners in HCI, the future – envisioning what it might be and questioning what it *should* be – are core concerns. For those focused on issues of sustainability, ethical considerations (the *should*) related to the future are recognized as "contested, situated and ever-changing" (Meyers & Nathan, 2016, p. 222). Sustainability-grounded, future-oriented design inquiries are developing as a signature strength of the ACM SIGCHI Conference on

Designing Interactive Systems (DIS), for example Hauser et al. (2014); Jeremijenko (2016); and McKinnon (2016). Particularly relevant to SHCI are projects that consider what and how people design and enact their designs while going about their daily lives.

Expanding the everyday designer scholarship beyond an examination of specific tools and their associated practices, Desjardins, Wang, and Wakkary (Chapter 11) introduce us to the idea of sustainable placemaking. They trace an appreciation for place-based, community-centered design approaches to the early work of Jane Jacobs and William Whyte, individuals who advocated for community-based urban planning 50 years ago (Project for Public Spaces, 2017). The authors apply ideas of placemaking to inform their analysis of a project with community gardeners in the Pacific Northwest and an autobiographical description of converting a camper van interior. The authors consider ways that everyday designers not only adapt objects, but also determinedly transform their immediate environments to better support what they hold as important. As a step away from focusing on the potential for interactive technologies (that others have designed) to support conceptualizations of sustainability, Desjardins, Wang, and Wakkary reflect on the work everyday designers (i.e., not professionals) undertake when adapting their immediate environments to better support their ideals, what they want to sustain in their daily lives. Through insights gained from their case studies, the authors demonstrate how themes of longevity, unfinishedness, and the multiplicity of everyday design strategies have much to offer future sustainable HCI and interaction design research.

Sharing an interest in humanity's resourcefulness and design sensibilities while rejecting a singular vision of a sustainable future, Bendor (Chapter 12) advocates for a pluralistic approach to sustainability inquiries. Drawing upon quotes from diverse personalities – from Pope Francis to Victor Papanek – Bendor encourages SHCI researchers to interrogate dominant conceptualizations of 'the future'. He asks us to expand our notions of what is possible and to imagine how the design of interactive technology can (indeed already has) help(ed) people envision and enact a multiplicity of futures. He promotes ideas of futurescaping and worldmaking as concepts that can encourage broader publics to actively participate in visioning activities, expanding what they understand as possible, rather than slipping into a mindset that accepts the future as predetermined and simply unfolding. Exploring these how these concepts play out in practice, Bendor reflects on the design decisions that informed a recent large-scale, multi-year, and futures-oriented media project.

The opportunity to be an active participant – envisioning and enacting an alternate world, a new *home* – is part of the allure of the 30-year-old Burning Man phenomenon. Tsaasan and Nardi (Chapter 13) offer those uninitiated to Burning Man an opportunity to learn about its history, practices, and values. The first author's decades of experience participating in Burning Man activities inform the understandings in the chapter. Although digital technologies are discussed in the chapter, the authors take care to help the reader comprehend the possibilities that lie in connecting the guiding principles and ethos of Burning Man with future SCHI research.

Strengers' response to the ideas expressed in Part 4 begins by acknowledging the shared sense of frustration many feel concerning the narrowness (in scope and perspective) of the SHCI field to date. She quickly moves on to demonstrate how the three chapters in this final section of the text provide exemplars of SHCI scholarship moving towards more diverse, participatory, and inspiring futures. Strengers' reflection develops an appreciation for fresh orientations, originality, and ingenuity, key characteristics of the authors who contributed chapters to Part 4.

Conclusion

In closing, we welcome you to this assemblage of reflections on sustainable human-computer interaction. We also have a charge for you: a responsibility. While you are reading, we ask that you consider these lines from Section 1 of the "General moral principles" laid out in the ACM Code of Ethics and Professional Conduct. It states, "[H]uman well-being requires a safe natural environment. Therefore, computing professionals should be alert to, and make others aware of, any potential harm to the local or global environment" (ACM, 2018). We ask all readers, whether you read one chapter or the entire text, to envision how insights from this collection can inform, deepen, and strengthen this statement. Perhaps rather than alerting others of potential harms, your contributions will inform possibilities for humans (and other beings) to flourish. In reference to the Burning Man slogan "No Spectators" (see Chapter 13), we request your participation. We challenge you to enact your vision of SHCI, with others, and with humility.

Notes

1 Photographer Chris Jordan is an exemplar of someone working to make the magnitude of toxic waste and its impacts comprehensible through still images, video and computer graphics. Jordan graciously allowed his work to be shown during the first Sustainability Special Interest Group meeting at the Conference of Human Factors in Computing in 2007. http://chrisjordan.com
2 www.sigchi.org/communities/hci-sustainability
3 http://acmlimits.org

References

Association for Computing Machinery (ACM). (2018). *ACM code of ethics and professional conduct*. Retrieved from https://ethics.acm.org

Bidwell, N.J., Siya, M., Marsden, G., Tucker, W.D., Tshemese, M., Gaven, N., Ntlangano, S., Robinson, S., & Eglinton, K.A. (2013). Walking and the social life of solar charging in rural africa. *ACM Transactions on Computer-Human Interaction 20*(4), article 22. https://doi.org/10.1145/2493524

Blevis, E. (2007). Sustainable interaction design: Invention & disposal, renewal & reuse. In *Proceedings of the SIGCHI Conference on Human Factors in Computing Systems* (pp. 503–512). New York: ACM. http://dx.doi.org/10.1145/1240624.1240705

Blevis, E., Bødker, S., Flach, J., Forlizzi, J., Jung, H., Kaptelinin, V., . . . Rizzo, A. (2015). Ecological perspectives in HCI: Promise, problems, and potential. In *Proceedings of the 33rd Annual ACM Conference Extended Abstracts on Human Factors in Computing Systems* (pp. 2401–2404). New York: ACM. https://doi.org/10.1145/2702613.2702634

Busse, D., Blevis, E., Beckwith, R., Bardzell, S., Sengers, P., Tomlinson, B., Nathan, L. P., & Mann, S. (2012). Social sustainability: An HCI agenda. In *CHI '12 Extended Abstracts on Human Factors in Computing Systems* (pp. 1151–1154). New York: ACM. https://doi.org/10.1145/2212776.2212409

DiSalvo, C., Sengers, P., & Brynjarsdóttir, H. (2010). Mapping the landscape of sustainable HCI. In *Proceedings of the SIGCHI Conference on Human Factors in Computing Systems* (pp. 1975–1984). New York: ACM. http://dx.doi.org/10.1145/1753326.1753625

Hauser, S., Desjardins, A., & Wakkary, R. (2014). Sfuture: Envisioning a sustainable university campus in 2065. In *Proceedings of the 2014 Companion Publication on Designing Interactive Systems* (pp. 29–32). New York: ACM. https://doi.org/10.1145/2598784.2602774

Hazas, M., Morley, J., Bates, O., & Friday, A. (2016). Are there limits to growth in data traffic? On time use, data generation and speed. In *Proceedings of the Second Workshop on Computing Within Limits*. New York: ACM. https://doi.org/10.1145/2926676.2926690

Hilty, L. M., & Aebischer, B. (Eds.). (2015). ICT innovations for sustainability. In *Advances in intelligent systems and computing* (Vol. 310). Springer.

Huh, J., Nathan, L. P., Silberman, E., Blevis, E., Tomlinson, B., Sengers, P., & Busse, D. (2010). Examining appropriation, re-use, and maintenance for sustainability. In *CHI '10 Extended Abstracts on Human Factors in Computing Systems* (pp. 4457–4460). New York: ACM. https://doi.org/10.1145/1753846.1754173

Jeremijenko, N. (2016). Creative agency and the space race of the 21st century: Towards a museum of natural futures. In *Proceedings of the 2016 ACM Conference on Designing Interactive Systems* (pp. 3–4). New York: ACM. https://doi.org/10.1145/2901790.2915254

Kitchin, R., & Dodge, M. (2011). *Code/space: Software and everyday life*. Cambridge, MA: MIT Press.

Knowles, B., Clear, A. K., Mann, S., Blevis, E., & Håkansson, M. (2016). Design patterns, principles, and strategies for Sustainable HCI. In *Proceedings of the 2016 CHI Conference Extended Abstracts on Human Factors in Computing Systems* (pp. 3581–3588). New York: ACM. https://doi.org/10.1145/2851581.2856497

Knowles, B., & Håkansson, M. (2016). A sustainable HCI knowledge base in progress. *Interactions, 23*(3), 74–76. https://doi.org/10.1145/2904896

Mankoff, J. C., Blevis, E., Borning, A., Friedman, B., Fussell, S. R., Hasbrouck, J., Woodruff, A., & Sengers, P. (2007). Environmental sustainability and interaction. In *CHI '07 Extended Abstracts on Human Factors in Computing Systems* (pp. 2121–2124). New York: ACM. http://dx.doi.org/10.1145/1240866.1240963

McKinnon, H. (2016). The [everyday] future by design: Opportunities for the design exploration of everyday sustainability. In *Proceedings of the 2016 ACM Conference Companion Publication on Designing Interactive Systems* (pp. 37–38). New York: ACM. https://doi.org/10.1145/2908805.2909424

Meyers, E. M., & Nathan, L. P. (2016). Impoverished visions of sustainability: Encouraging disruption in digital learning environments. In *Proceedings of the 19th ACM Conference on Computer-Supported Cooperative Work & Social Computing* (pp. 222–232). New York: ACM. https://doi.org/10.1145/2818048.2819987

Olopade, D. (2014). *The bright continent: Breaking rules and making change in modern Africa*. Boston: Houghton Mifflin Harcourt.

Preist, C., Schien, D., & Blevis, E. (2016). Understanding and mitigating the effects of device and cloud service design decisions on the environmental footprint of digital infrastructure. In *Proceedings of the 2016 CHI Conference on Human Factors in Computing Systems* (pp. 1324–1337). New York: ACM. https://doi.org/10.1145/2858036.2858378

Pierce, J., Schiano, D. J., & Paulos, E. (2010). Home, habits, and energy: Examining domestic interactions and energy consumption. In *Proceedings of the SIGCHI Conference on Human Factors in Computing Systems* (pp. 1985–1994). New York: ACM. http://dx.doi.org/10.1145/1753326.1753627

Project for Public Spaces. (2017). *What is placemaking?* Retrieved from www.pps.org/reference/what_is_placemaking/

Silberman, M. S., Blevis, E., Huang, E., Nardi, B. A., Nathan, L. P., Busse, D., . . . Mann, S. (2014a). What have we learned? A SIGCHI HCI & sustainability community workshop. In *CHI '14 Extended Abstracts on Human Factors in Computing Systems* (pp. 143–146). New York: ACM. https://doi.org/10.1145/2559206.2559238

Silberman, M. S., Nathan, L., Knowles, B., Bendor, R., Clear, A., Håkansson, M., . . . Mankoff, J. (2014b). Next steps for sustainable HCI. *Interactions, 21*(5), 66–69. https://doi.org/10.1145/2651820

Strengers, Y. (2011). Designing eco-feedback systems for everyday life. In *Proceedings of the SIGCHI Conference on Human Factors in Computing Systems* (pp. 2135–2144). New York: ACM. https://doi.org/10.1145/1978942.1979252

Tomlinson, B., Silberman, M. S., Patterson, D., Pan, Y., & Blevis, E. (2012). Collapse informatics: Augmenting the sustainability & ICT4D discourse in HCI. In *Proceedings of the SIGCHI Conference on Human Factors in Computing Systems* (pp. 655–664). New York: ACM. http://dx.doi.org/10.1145/2207676.2207770

Photo Essay 2

Artifice and Nature (2017). Bird of paradise plants in the foreground contrast with the ongoing construction of the built environment in the background of this image. The sustainable interaction design principle of "*using natural models and reflection*" is implicated in many of the chapters, and yet finding inspiration in the design and use of artifice in the model of Nature's zero-waste cycles of life continues to elude. Reflection on Remy: *Communicating SHCI Research to Practitioners and Stakeholders*; and Desjardins: *Reflections on longevity, unfinishedness and design-in-living*; and Roedl, Odom, and Blevis: *Three Principles of Sustainable Interaction Design, Revisited*.

Eli Blevis

Part 1
Assessing the field

1 Three principles of sustainable interaction design, revisited

David Roedl, William Odom, and Eli Blevis

Introduction

In a 2007 paper, Blevis introduced the notion of sustainable interaction design (SID), "the perspective that sustainability can and should be a central focus of interaction design" (p. 503). In the years since, sustainability has indeed developed into a significant (if not yet central) theme for research within the HCI and interaction design communities. This body of work now encompasses a wide variety of topics, theories, and approaches (see DiSalvo et al., 2010, for a critical survey). Although Blevis is widely cited for bringing sustainability to the fore of HCI discourse, the specific concepts presented in his 2007 paper have been less frequently referenced and discussed. In this chapter, we revisit Blevis' original framing of SID – in particular his five principles – and review the ways these concepts have been explored and further developed in scholarly literature.

Blevis' (2007) paper explicitly focuses on environmental sustainability, noting that environmental concerns can be approached "both from the point of view of how interactive technologies can be used to promote more sustainable behaviors and – with more emphasis here – from the point of view of how sustainability can be applied as a critical lens to the design of interactive systems, themselves" (ibid). Although the former approach has since been embraced by a large genre of research in HCI (e.g., persuasive sustainability), Blevis' paper deals almost exclusively with the latter. In particular, Blevis proposes a set of conceptual tools (a rubric and five principles) for understanding and mitigating the material effects that interaction designs can have on the environment. By material effects, Blevis calls attention to the environmental consequences that result throughout the lifecycles of digital technologies, including energy consumption, material resource use, and waste production.

Blevis presents a Rubric of Material Effects (RoME) for assessing the potential material outcomes that result from particular design choices. The items are listed in approximate order of least to most desirable from a sustainability perspective: *disposal, salvage, recycling, remanufacturing for reuse, reuse as is, achieving longevity of use, sharing for maximal use, achieving heirloom status, finding wholesome alternatives to use, active repair of misuse.* As explained in (Blevis, 2006), the rubric is intended to be used both as a tool of *design criticism*, which provides "the understandings needed to uncover the effects of present courses of action and inform future ones," and *critical design* – "the actual practice of design with the materials of information technologies critical to the goal of promoting sustainable ways of being" (p. 1).[1] Following the rubric, Blevis (2007) presents five design principles, which he describes as informal rules or goals for achieving less harmful material outcomes in design: (1) *linking invention & disposal*, (2) *promoting renewal & reuse*, (3) *promoting quality & equality*, (4) *decoupling ownership & identity*, and (5) *using natural models & reflection*. Blevis states that the first two principles are targeted at "considering how the use of digital materials actually prompts the use of physical ones," while the latter three principles "all relate to finding ways to promote renewal & reuse over invention & disposal" (2007, p. 507).

While research in the area of sustainable HCI engages with a wide variety of social and environmental issues, Blevis' original paper is primarily focused on developing a foundation for design that can mitigate the environmental impacts of digital technologies themselves. In this retrospective chapter we review the current state of knowledge regarding this challenge. In particular, we ask: How have the research goals implied by Blevis' principles been pursued and developed? What progress has been made in developing knowledge that is relevant and actionable for design? What challenges and open questions remain? For each of the first three principles,[2] we summarize scholarly contributions found in the literature,[3] and synthesize the key insights that have been learned. In addition, we identify gaps in the knowledge and suggest opportunities for new research. Finally, in the concluding section, we reflect on the extent that the principles remain relevant to the contemporary and future challenges of designing digital technologies in a manner compatible with environmental sustainability.

Linking invention & disposal

Blevis defines the first principle of SID as: "the idea that any design of new objects or systems . . . is incomplete without a corresponding account of what will become of the objects or systems that are displaced or obsoleted by such inventions" (Blevis, 2007, p. 503). In his discussion, Blevis calls attention to the ways that new designs contribute to premature obsolescence and disposal, often as part of intentional and coordinated business strategies. Through criticism of particular cases, such as Apple's iPod, he observes that obsolescence is achieved through regular reinvention of product functionality, form, and style in concert with fashion trends and marketing tactics. Importantly, Blevis argues that software design is equally implicated in the process of obsolescence. While software has the potential to reduce material effects by updating the uses of existing hardware, more often the opposite occurs as software releases drive demand for new hardware, and cause premature disposal of existing devices.

What we have learned

Researchers have since elaborated on the various ways HCI and interaction design contribute to obsolescence, both at the levels of practice and theory. Several empirical studies have revealed the industry practices and other social and contextual factors that influence repetitive consumption of digital products. For example, Hanks et al. (2008) found millenials prefer newness over value, quality, and durability when purchasing digital devices. Huang and Truong (2008a, b) found that rapid turnover in mobile phones is driven by a combination of marketing practices: the introduction of new styles, functionality, and most importantly, contract-based pricing schemes. Other scholars have pointed out that techniques of planned obsolescence have a long history within corporate design practice (e.g., Huh & Ackerman, 2009, following Sterne, 2007). Despite this fact, the economic imperative towards obsolescence is not often acknowledged or discussed within HCI literature.

Recently, increased environmental awareness has prompted reflection about how the logic of invention and disposal is embedded in the theory and methods of design. For example, Pierce (2012, p. 959) draws on philosopher Tony Fry to argue the inherent destructiveness of design is usually obscured in favor of celebrating what is created. Moreover, a growing collection of critical voices[4] suggests that HCI's framing of users as consumers and its disciplinary demand for innovation help to advance an ideology of consumerism. This orientation tends to result in problematic 'solutions' to sustainability that involve creating more and more products, rather than maintaining existing devices or eliminating waste.

Opportunities for future research

In order to account for the ways that invention may cause disposal, researchers need to attend to the various forces that drive obsolescence. As shown in the studies mentioned above, these forces are numerous and complex. They include the interaction between hardware and software updates, product styles, fashion trends, advertising, contract and pricing schemes. They also include the underling economic ideology within which design is intertwined and that may be difficult to see. As Blevis has argued, criticism of particular cases is an important method of research for dealing with this complexity. It is important to note that the type of criticism called for here is one that goes beyond HCI's traditional focus on individual humans and computers. This point resonates with recent work in HCI that decenters the individual user to study social phenomena at a larger scale (Pierce et al., 2013; Dourish, 2010). This move also underlines the need for transdisciplinary thinking that may draw on expertise from fields such as design, STS, political economy, cultural anthropology, sociology, and cultural studies. While the need for such criticism has been argued, and some sources for theoretical tools have been identified, examples to date are quite rare within HCI. The discourse is overwhelmingly focused on what design creates, rather than what it displaces and destroys.

In particular, we note a lack of rigorous engagement with the political economic dimensions of obsolescence. Can capitalist economies survive without planned obsolescence, and if so, how? Must the computing industry pursue Moore's law indefinitely, or is there a point at which it can sustain itself while focusing on alternative engineering goals? These questions represent a significant knowledge gap and space of opportunity for future research. Nardi and Ekbia have called political economy the "elephant in the room" for HCI (2015). In this volume, Nardi and Ekbia outline practical strategies to help researchers begin to "grapple with the economy as the underlying mechanism implicated in the predicament [of unsustainability]".

Even with the ability to identify and understand the dynamics of obsolescence, how may designers and researchers avoid contributing to cycles of invention and disposal? This begins with first considering what problems actually merit a designer's (or researcher's) attention, as argued by Knouf (2009) following Papanek (1971). Instead of creating new commodities and tools for consumption, interaction designers might instead devote their efforts to designing technologies that address basic human needs, as in the field of HCI4D and Collapse Informatics (Tomlinson et al., 2013). Or, designers might choose not to design any new things, but rather to practice strategies of *elimination design* (Pierce, 2012; Fry, 2009). As suggested by Pierce (2012) this might entail a shift away from creating products[5] towards methods of design argumentation and provocation, such as critical design (Dunne & Raby, 2001; Bardzell & Bardzell, 2013), visual thinking and image making (Blevis, 2011). Alternatively, designers might focus on enabling the renewal and reuse of existing technology, as we discuss in the following section.

Promoting renewal & reuse

Blevis defines the second principle of SID as the idea that any design should "first and foremost consider the possibilities for renewal & reuse of existing objects or systems" (2007, p. 503). Blevis states that this principle relates to several of the categories from the RoME, including *salvage, recycling, remanufacturing for reuse, reuse as is*, and *sharing for maximal use*. In his discussion, Blevis re-iterates that software as a digital material seems to provide opportunities for the renewal and reuse of existing hardware, but that in practice the opposite typically occurs as new software prompts the invention of new hardware and vice versa.

What we have learned

Since the introduction of the SID principles, a robust body of empirical research[6] has emerged exploring the various ways that users respond to moments of breakdown and obsolescence with digital devices. This work has revealed numerous barriers that currently inhibit renewal and reuse, including: lack of knowledge and tools, restrictive and opaque hardware and software, service contracts, copyright protection, and corporate and organizational policies that limit user access. In a more insidious fashion, the support of renewal and reuse practices is also marginalized by dominate industry rhetoric that celebrates design and passive use, while ignoring the myriad creative activities required to sustain technology throughout its lifespan (Jackson et al., 2012).

In spite of these barriers, research has shown that many people actively labor to extend the life of technological objects across a variety of contexts. Such practices include transfer of ownership, maintenance, creative appropriation and repurposing, and repair. A recurring theme of the research is that these efforts frequently require significant amounts of time, creativity, and skill. Research into renewal and reuse has coincided and overlapped with a growing HCI interest in various creative practices that transcend passive use, including crafts, appropriation, hacking, everyday design, DIY and maker cultures (Buechley et al., 2009). Collectively, these works represent a movement in HCI from designing fixed products for passive consumption towards the design of flexible tools and resources for creative appropriation and adaptation (Roedl et al., 2015).

The heterogeneity of renewal practices demonstrates that there is room for finer distinctions of vocabulary to be made among Blevis' RoME. For example, within the category of *reuse as is*, we can talk about transfer of ownership through various processes of dispossession and reacquisition (Pierce & Paulos, 2011). On the other hand, creative acts of maintenance, appropriation, repurposing, and everyday design are forms of renewal that blur the boundaries between *salvage, remanufacturing for reuse*, and *reuse as is*. These acts renew the meaning and value of an artifact (through sometimes with minimal changes to its physical form), allowing for continued use by either a single owner or across multiple owners. In addition, repair has been shown to be an important category of renewal (although not specifically mentioned in RoME) that enables continued use for a current or secondary owner. Successful repairs often depend on the salvage of spare parts and materials from previously discarded devices. Repairs can vary widely in sophistication from simple fixes to difficult projects that require expert hands, tools, and diagnostic knowledge. Moreover, all these practices of renewal have shown to take place with various degrees of formalization and social arrangement, ranging from casual reuse to hobbyist DIY projects to entrepreneurial craftwork, etc. Of course many of these practices overlap and intermingle. The goal in describing heterogeneity with new vocabulary is not so much to draw distinct lines as to give proper consideration to a wide space of technological activity that has until recently been understudied.

Another important contribution of this research has been to demonstrate that successful renewal and reuse depend as much, if not more, on social and cultural factors as on the design of artifacts.[7] First and foremost, a device must be perceived as worth fixing, updating, or reacquiring. This depends somewhat on the material and aesthetic quality of the artifact (as we discuss in the following section), but also on perceptions of worth that vary (and are contested) among people and cultures, as well as on the economic exchange value of devices, replacement parts, and labor. Second, the likelihood for renewal and reuse depend significantly on the availability of knowledge, skill, tools, and social support. Finally there is a significant political dimension affecting the possibility of renewal and reuse. By this we refer to the legal, economic, and social parameters that designate who is permitted to modify, repair, or resell technology. Who is deemed responsible or privileged to do this work, and how is such work esteemed and compensated by society? Although more research is clearly needed to understand these issues, we can state that, as a general implication, designing for renewal and reuse requires a strategic and systemic approach, accounting for social and economic infrastructures as much as artifacts.

Opportunities for future research

While the scholarship on renewal and reuse is growing, and the above contributions are significant, we believe this area is still under-researched compared to design's long preoccupation with innovation that drives new production. Research opportunities abound both to better understand current practices of renewal and reuse and to identify interventions that can help support them. Here we point to just a few exciting questions for future work.

First, much of above research has focused on the repair or repurposing of hardware, but relatively less attention has been given to the effects of software. In what ways can software be designed to enable the reuse of existing hardware? More examples and principles towards this goal are needed.

Second, numerous authors have pointed to the limiting effects of restrictive, "black-box" design, and have argued that design principles of openness and transparency can better facilitate renewal. We believe the connection between openness and sustainability is a significant opportunity space, and traditions of thought among Free and Open Source Software communities may be a useful source of inspiration. In addition, given the growing popularity of DIY and maker practices, will companies begin to view modular hardware as a more viable design strategy?

Finally, recent research has mentioned economic challenges facing renewal and reuse, but these issues have yet to be explored in depth. For example, is it possible for repair and reuse work to be adequately compensated and sustained in the face of competition from cheap, disposable production? Are there new business models or regulatory policies that can address these problems?

Promoting quality and equality

The principle of promoting quality and equality is originally stated in (Blevis, 2007) and emphasizes the imperative to design technologies with superior materials that promote enduring use and care. The broader aim of this principle is to support the transfer of ownership over time and, in so doing, reduce social and environmental consequences for the collective human condition that come with the disposal of digital technology.

What we have learned

The principle of promoting quality and equality has since been further explored and developed in HCI literature. Many of these works aim to develop the HCI community's understanding of strategies for extending the lifespan of interactive technologies. The predominant focus has been on translating theoretical insights related to theories of attachment into recommendations for design practice. A less prevalent yet important area of work has focused on designing radical responses to contemporary consumer technologies to explore how a deeper sense of care for interactive systems might be engendered.

In 2007, Blevis and Stolterman propose the use of personal inventories as a field method to elicit and record designerly insights into the kinds of objects people do (and don't) become attached to over time. Blevis and Stolterman extend the connection between quality and equality to the work of contemporary philosopher of technology Peter-Paul Verbeek (2005), who argues for critically analyzing relations between humans and artifacts as a means to understand issues shaping unsustainable practices. Inspired by Verbeek, Blevis, and Stotlerman describe a pilot study that began to develop the personal inventories method as a means to record and analyze factors that shaped people's relations to particular domestic artifacts (digital and non-digital). In 2009, Odom et al. present a more developed framework for categorizing and analyzing human-product relationships in the context of sustainable interaction design. Findings from personal inventory studies with 32 households revealed that *engagement, histories, augmentation*, and *perceived durability* were four main dimensions of human-product relations that shaped attachment. The dimensions of Odom et al.'s attachment framework can be seen as strategies for pursing quality and equality in that they articulate ways in which a design object can acquire histories and narratives as it evolves over time.

In parallel to this work, Wakkary and Tanenbaum (2009) explore the notion of quality and equality through the lens of everyday design (Wakkary & Maestri, 2007). Through in-depth ethnographic observations of families' everyday practices, the authors surface several exemplars of the notion of quality and equality in everyday household objects. The most striking exemplar is Kerry's recipe book that, despite being over a decade old, continues to acquire new recipes, newspaper clippings, handwritten annotations, personal lists, etc. (see Wakkary & Tanenbaum, 2009, p. 370). It is a clear example of an artifact that has achieved heirloom status.

Jung et al. (2011) extends these collective works further by focusing specially on collecting "deep narratives" of how artifacts achieved heirloom status. Specifically, they highlight three nuanced pathways toward this goal: *intimacy is accumulated over time, investment of concerted effort to learn and control sophisticated functionality*, and *the implicit emergence of values across sets or ensembles of artifacts*. The notion that attachment can arise through the investment of effort to develop expertise in the operation of a technical artifact also emerges in Huh et al.'s study of people's long-term use of PDA devices (2010).

Gegenbauer and Huang (2012a, b) later conducted a set of personal inventory studies with Swiss households that aimed to extend these works and, in particular, Odom et al.'s (2009) initial attachment framework to include three additional categories to achieve longevity of use: *earned functionality, perceived worth*, and *sufficiency*. This corpus of work and the studies mentioned previously offer a more developed foundation to direct generative design research investigations into notions of quality and equality.

Opportunities for future research

Through tracing the proposal and emergence of the principle of quality and equality it becomes clear that the issues shaping human relationships with technological artifacts are of increasing concern in the HCI community. Many of the reviewed studies examine the experiential nature of people's attachments to well-loved non-digital artifacts as a lens to understand factors contributing to digital things falling short of achieving longevity of use. Following Tony Fry's notion that redirective practitioners must "dig where they stand" (Fry, 2009), it is understandable that researchers largely based these studies within the local contexts that they themselves live and work. As a result nearly all of the known work at the intersection of product attachment and sustainable HCI design has been conducted within North America or Western Europe. There is an opportunity to expand research involving personal inventories to non-western contexts. It may be the case that notions of sentimentality, personal narrative, and attachment that we see so clearly in human-product relations within western contexts may not generalize to other areas of the world. Expanding our understanding of the multiplicity of social, cultural, political, economic, and even spiritual dimensions that shape why people preserve some things and discard others in the context of HCI may offer new and diverse pathways for design to play a role in facilitating more sustainable ways of being.

Additionally, within the majority of the reviewed works there has been a special emphasis on developing new insights, tactics, and strategies to redirect professional design practice toward creating interactive technologies of such a high quality that they may be held onto and develop a richness and value associated with heirloom artifacts. However, it also was clear that design frameworks, such as the attachment framework, that are surfaced from empirical research and ongoing critical reflection on the collected phenomena can themselves fall short of being directly applicable to designers. Clearly, we need better ways of influencing design practice. One approach, as suggested by Remy et al. (2015), could be even further constraining HCI design framework so as to minimize any ambiguity, which could disrupt an otherwise straightforward prescriptive set of guidelines for practitioners. An alternative research direction could focus on creating and implementing new conceptual design artifacts that themselves operate as embodied theoretical articulations of particular aspects of the attachment framework or the principle of quality and equality. As HCI continues to embrace more reflective forms of knowledge production (e.g., Zimmerman et al., 2007; Stolterman & Wiberg, 2010), this could be a viable way forward.

While the concept-driven direction mentioned directly above has much promise, it also comes with unresolved issues. Is it possible to create enduring computational artifacts when the digital materials comprising them will still become obsolete over relatively short amounts of time? These deeper issues of quality hardware design, implementation, and manufacturing introduce unavoidable tensions that often go unacknowledged.

Interestingly, recent work has proposed that the HCI community ought to invert this model completely. Holmer et al. (2014) argue that one problem might lie in the assumption that designers are the experts in how heirloom quality might be crafted or achieved. Instead they propose: "What would it be like to re-frame the making of heirloom systems from an activity primarily done by experts (designers, artists, artisans, and other craft people) to an endeavor of participatory making open to non-experts as well?" (2014, p. 30). Indeed, this seems like an intriguing framing for future research and could help expand the principle of quality and equality itself. Indeed, projects in the commercial sector, such as the Fairphone (www.fairphone.com), offer new examples of how ethically sourced materials, labor, and longer-lasting design can be productively integrated as central concerns in emerging technology products. The Fairphone is discussed and studied in the context of HCI in Joshi and Cerratto-Pargman (2015). Projects like Fairphone are not without criticism – the issue that is sometimes raised is that to scale well, we also need to find ways to make heirloom quality be an affordable alternative, whereas the Fairphone is relatively expensive.

Finally, while the notion of quality resonated across all works reviewed in this section, the notion of equality was often implicit or assumed. In other words, if we are able to design a computational object to achieve heirloom status, then it is often taken for granted that it would be passed down to a future generation or otherwise to someone else that would continue to appreciate this. Here, we can consider equality as the successful ways in which quality is maintained in a computational object and how that thing is eventually transferred. There is a major need for more research to understand the tensions and complications a computational object will encounter as quality is maintained and ownership and possession is transferred over time.

Conclusion

As evident in our review, the first three principles of SID have helped to frame and inspire a robust and evolving space of design research. Rather than offering simple prescriptions, the SID principles articulate ambitious ethical goals that have been used both to evaluate current interactions and to explore possible design alternatives. Research over the last 10 years has yielded considerable insight into both the challenges and opportunities that exist for creating long-lasting, environmentally sustainable technologies. Still, it is clear that there is a long way to go towards achieving this goal, and the global rate of consumption for digital devices continues to accelerate. We hope that the research opportunities identified above, together with the many contributions found in this volume, can continue to scaffold knowledge that is critical for reversing this trend. Moreover, the two SID principles not covered in this chapter, *decoupling ownership & identity* and *using natural models & reflection*, represent additional spaces of opportunity to be explored in future work.

Notes

1 It should be noted that Blevis' notion of critical design is a different and more general category than that defined by Dunne and Raby (2001) and discussed by Bardzell and Bardzell (2013).

2 Due to limitations of space, we focus our discussion on just the first three principles, which are the ones that have received the most attention to date. We also discuss some items from the Rubric of Material Effects (RoME) as they come up in relevance to each principle.
3 Our archive is a collection of 230 papers that reference the 2007 paper by Blevis, which we found via Google Scholar in February of 2015. The majority of these works are published in HCI venues such as the proceedings for ACM conferences CHI, DIS, Ubicomp, etc. While outside of our corpus of literature, the recent work by Preist et al. (2016) offers a substantial extension of the design rubric in Blevis' seminal (2007) paper. Their design-critical lens examines the complex link between devices and cloud service decisions, the resulting consequences for the human condition, and the imperative for radical change beyond the "eco-efficiency" paradigm. In this recent work (a) linking invention and disposal is associated with linking infrastructural expansion and obsolescence, (b) promoting renewal and reuse is associated with promoting infrastructural use-efficiency and sharing, and (c) promoting quality and equality is associated with promoting reliable infrastructure from sustainable sources and promoting equitable distribution of bandwith. These correspondences and other frames are tabled as well in Blevis et al. (2017).
4 See, for example: Knouf (2009), Wakkary and Tanenbaum (2009), Goodman (2009), Fallman (2009, p. 58), Schweikardt (2009), Dourish (2010), Brynjarsdóttir et al. (2012), Hakansson and Sengers, (2013, p. 2728).
5 Such a shift speaks directly to Remy and Huang's (this volume) discussion of rethinking both the target audience and framing of HCI research output.
6 See, for example: Huang & Truong (2008a, b), Hunag et al. (2009), Barba (2008), Huh & Ackerman (2009), Huh et al. (2010), Wakkary & Tannenbaum (2009), Wakkary et al. (2013), Hasbrouck (2011), Pierce and Paulos (2011), Kim and Paulos (2011), Zhang and Wakkary (2011), Maestri and Wakkary (2011), Vyas (2012), Jackson et al. (2012, 2014), Eriksson et al. (2014), Rosner and Ames (2014).
7 This theme has come up in many studies, but is perhaps shown most clearly in the work of Jackson et al. (2012, 2014) and Rosner and Ames (2014), who have explored the social complexity of repair practices through ethnographic field study in Namibia, Bangladesh, Paraguay, and California.

References

Barba, E. (2008). Getting mod: A look at modularity in mobile systems. In *Proceedings of the 9th workshop on Mobile Computing Systems and Applications*. http://doi.org/10.1145/1411759.1411766

Bardzell, J., & Bardzell, S. (2013). What is "critical" about critical design? In *Proceedings of the SIGCHI Conference on Human Factors in Computing Systems*. New York: ACM. http://doi.org/10.1145/2470654.2466451

Blevis, E. (2006). Advancing sustainable interaction design: Two perspectives on material effects. *Design Philosophy Papers*, 4(4), 209–230. doi: 10.2752/144871306X13966268131875

Blevis, E. (2007). Sustainable interaction design: Invention & disposal, renewal & reuse. In *Proceedings of the SIGCHI Conference on Human Factors in Computing Systems*. New York: ACM. http://doi.org/10.1145/1240624.1240705

Blevis, E. (2011). Digital imagery as meaning and form in HCI and design: An introduction to the Visual Thinking Backpage Gallery. *Interactions*, 18(5). http://doi.org/10.1145/2008176.2008190

Blevis, E., Preist, C., Schien, D., & Ho, P. (2017). Further connecting sustainable interaction design with sustainable digital infrastructure design. In *Proceedings of ACM Limits 2017*. Santa Barbara: ACM.

Blevis, E., & Stolterman, E. (2007). Ensoulment and sustainable interaction design. In *Proceedings for the International Association of Societies of Design Research (IASDR)*.

Brynjarsdottir, H., Hakansson, M., Pierce, J., Baumer, E., DiSalvo, C., & Sengers, P. (2012). Sustainably unpersuaded: How persuasion narrows our vision of sustainability. In *Proceedings of the SIGCHI Conference on Human Factors in Computing Systems*. New York: ACM. http://doi.org/10.1145/2207676.2208539

Buechley, L., Rosner, D. K., Paulos, E., & Williams, A. (2009). DIY for CHI: methods, communities, and values of reuse and customization. In *CHI '09 Extended Abstracts on Human Factors in Computing Systems (CHI EA '09)* (pp. 4823–4826). New York: ACM. DOI: https://doi.org/10.1145/1520340.1520750

DiSalvo, C., Sengers, P., & Brynjarsdottir, H. (2010). Mapping the landscape of sustainable HCI. In *Proceedings of the SIGCHI Conference on Human Factors in Computing Systems*. New York: ACM. http://doi.org/10.1145/1753326.1753625

Dourish, P. (2010). HCI and environmental sustainability: The politics of design and the design of politics. In *Proceedings of the 8th ACM Conference on Designing Interactive Systems*. New York: ACM. http://doi.org/10.1145/1858171.1858173

Dunne, A., & Raby, F. (2001). *Design Noir: The secret life of electronic objects*. Birkhäuser Basel.

Ekbia, H., & Nardi, B. (2015). The political economy of computing: The elephant in the HCI room. *Interactions* 22(6), 46–49. DOI: https://doi.org/10.1145/2832117

Eriksson, E., Pargman, D., & Artman, H. (2014). Usability as a threat to a sustainable future-Induced disability through better HCI. *Workshop Proceedings for "Sustainable HCI: What Have We Learned" at CHI 2014*, Toronto.

Fallman, D. (2009). A different way of seeing: Albert Borgmann's philosophy of technology and human – computer interaction. *AI & Society*, 25(1), 53–60. http://doi.org/10.1007/s00146-009-0234-1

Fry, T. (2009). *Design futuring*. London: Berg.

Gegenbauer, S., & Huang, E. M. (2012a). Inspiring the design of longer-lived electronics through an understanding of personal attachment. In *Proceedings of the Designing Interactive Systems Conference*. New York: ACM. http://doi.org/10.1145/2317956.2318052

Gegenbauer, S., & Huang, E. M. (2012b). iPods, Ataris, and Polaroids: A personal inventories study of out-of-use electronics in Swiss households. In *Proceedings of the 2012 ACM Conference on Ubiquitous Computing*. New York: ACM. http://doi.org/10.1145/2370216.2370294

Goodman, E. (2009). Three environmental discourses in human-computer interaction. *CHI '09 Extended Abstracts on Human Factors in Computing Systems*. http://doi.org/10.1145/1520340.1520358

Hakansson, M., & Sengers, P. (2013). Beyond being green: Simple living families and ICT. In *Proceedings of the SIGCHI Conference on Human Factors in Computing Systems*. New York: ACM. http://doi.org/10.1145/2470654.2481378

Hanks, K., Odom, W., Roedl, D., & Blevis, E. (2008). Sustainable millennials: Attitudes towards sustainability and the material effects of interactive technologies. In *Proceedings of the SIGCHI Conference on Human Factors in Computing Systems* (pp. 333–342). New York: ACM. http://doi.org/10.1145/1357054.1357111

Hasbrouck, J. (2011). From refurbish to renew: The trope of healing and sustainable technology design. *Design Philosophy Papers*, 9(1), 5–22.

Holmer, H. B., DiSalvo, C., Sengers, P., & Lodato, T. (2015). Constructing and constraining participation in participatory arts and HCI. *International Journal of Human Computer Studies*, 74(C), 107–123. https://doi.org/10.1016/j.ijhcs.2014.10.003

Huang, E. M., & Truong, K. N. (2008a). Breaking the disposable technology paradigm: opportunities for sustainable interaction design for mobile phones. In *Proceedings of the SIGCHI Conference on Human Factors in Computing Systems*. New York: ACM. http://doi.org/10.1145/1357054.1357110

Huang, E. M., & Truong, K. N. (2008b). SUSTAINABLY OURS: Situated sustainability for mobile phones. *Interactions*, 15(2), 16–19.

Huang, E. M., Yatani, K., Truong, K. N., Kientz, J. A., & Patel, S. N. (2009). Understanding mobile phone situated sustainability: The influence of local constraints and practices on transferability. *IEEE Pervasive Computing*, 8(1).

Huh, J., & Ackerman, M. S. (2009). Obsolescence: Uncovering values in technology use. *M/C Journal*, 12(3). http://doi.org/10.1145/2160673.2160702

Huh, J., Nam, K., & Sharma, N. (2010). Finding the lost treasure: Understanding reuse of used computing devices. In *Proceedings of the SIGCHI Conference on Human Factors in Computing Systems*. New York: ACM. http://doi.org/10.1145/1753326.1753607

Jackson, S. J., Ahmed, S. I., & Rifat, M. R. (2014). Learning, innovation, and sustainability among mobile phone repairers in Dhaka, Bangladesh. In *Proceedings of the 2014 Conference on Designing Interactive Systems* (pp. 905–991). New York: ACM. https://doi.org/10.1145/2598510.2598576

Jackson, S. J., Pompe, A., & Krieshok, G. (2012). Repair worlds: Maintenance, repair, and ICT for development in rural Namibia. In *Proceedings of the ACM 2012 conference on Computer Supported Cooperative Work*. New York: ACM. http://doi.org/10.1145/2145204.2145224

Joshi, S., & Cerratto Pargman, T. (2015). In search of fairness: Critical design alternatives for sustainability. In *Proceedings of The Fifth Decennial Aarhus Conference on Critical Alternatives* (AA '15) (pp. 37–40). Aarhus University Press.

Jung, H., Bardzell, S., Blevis, E., Pierce, J., & Stolterman, E. (2011). How deep is your love: Deep narratives of ensoulment and heirloom status. *International Journal of Design*, 5(1), 59–71.

Kim, S., & Paulos, E. (2011). Practices in the creative reuse of e-waste. In *Proceedings of the SIGCHI Conference on Human Factors in Computing Systems*. New York: ACM. http://doi.org/10.1145/1978942.1979292

Knouf, N. A. (2009). HCI for the real world. In *Extended Abstracts on Human Factors in Computing Systems*. New York: ACM. http://doi.org/10.1145/1520340.1520361

Maestri, L., & Wakkary, R. (2011). Understanding repair as a creative process of everyday design. In *Proceedings of the 8th ACM conference on Creativity and Cognition*. New York: ACM. http://doi.org/10.1145/2069618.2069633

Mankoff, J. C., Blevis, E., Borning, A., Friedman, B., Fussell, S. R., Hasbrouck, J., et al. (2007). Environmental sustainability and interaction. Extended Abstracts on Human Factors in Computing Systems, ACM. http://doi.org/10.1145/1240866.1240963

Odom, W. (2008). Personal inventories: toward durable human-product relationships. *Extended Abstracts on Human Factors in Computing Systems*, ACM, New York. http://doi.org/10.1145/1358628.1358929

Odom, W., Blevis, E., & Stolterman, E. (2008). SUSTAINABLY OURS: Personal inventories in the context of sustainability and interaction design. interactions, 15(5), 16–20, ACM.

Odom, W., Pierce, J., Stolterman, E., & Blevis, E. Understanding why we preserve some things and discard others in the context of interaction design. In *Proceedings of the SIGCHI Conference on Human Factors in Computing Systems* (pp. 1053–1062). New York: ACM. http://doi.org/10.1145/1518701.1518862

Odom, W., Pierce, J., Stolterman, E., & Blevis, E. (2009). Understanding why we preserve some things and discard others in the context of interaction design. In *Proceedings of the SIGCHI Conference on Human Factors in Computing Systems (CHI '09)* (1053–1062). New York: ACM. DOI: https://doi.org/10.1145/1518701.1518862

Papanek, V. (1971). *Design for the real world: Human ecology and social change*. New York: Pantheon Books.

Pierce, J. (2012). Undesigning technology: considering the negation of design by design. In *Proceedings of the SIGCHI Conference on Human Factors in Computing Systems*. New York: ACM. http://doi.org/10.1145/2207676.2208540

Pierce, J., & Paulos, E. (2011). Second-hand interactions: Investigating reacquisition and dispossession practices around domestic objects. In *Proceedings of the SIGCHI Conference on Human Factors in Computing Systems*. New York: ACM. http://doi.org/10.1145/1978942.1979291

Pierce, J., Strengers, Y., Sengers, P., & Bødker, S. (2013). Introduction to the special issue on practice-oriented approaches to sustainable HCI. *ACM Transactions on Computer-Human Interaction, 20*(4). http://doi.org/10.1145/2509404.2494260

Preist, C., Schien, D., & Blevis, E. (2016). Understanding and mitigating the effects of device and cloud service design decisions on the environmental footprint of digital infrastructure. In *Proceedings of the 34th Annual ACM Conference on Human Factors in Computing Systems (CHI'16)* (pp. 1324–1337). New York: ACM. https://doi.org/10.1145/2858036.2858378

Remy, C., Gegenbauer, S., & Huang, E. M. (2015). Bridging the theory-practice gap: Lessons and challenges of applying the attachment framework for sustainable HCI design. In *Proceedings of the 33rd Annual ACM Conference on Human Factors in Computing Systems* (pp. 1305–1314). New York: ACM. https://doi.org/10.1145/2702123.2702567

Roedl, D., Bardzell, S., & Bardzell, J. (2015). Sustainable making? Balancing optimism and criticism in HCI discourse. *ACM Transactions on Computer-Human Interaction, 22*(3), 1–27.

Rosner, D. K., & Ames, M. (2014). Designing for repair? Infrastructures and materialities of breakdown. In *Proceedings of the 17th ACM conference on Computer Supported Cooperative Work & Social Computing* (pp. 319–331). New York: ACM. http://doi.org/10.1145/2531602.2531692

Schweikardt, E. (2009). SUSTAINABLY OURS: User centered is off center. *Interactions, 16*(3). http://doi.org/10.1145/1516016.1516019

Sterne, J. (2007). Out with the trash: On the future of new media. *Residual Media*, 16–31.

Stolterman, E., & Wiberg, M. (2010). Concept-driven interaction design research. *Human–Computer Interaction, 25*(2), 95–118. https://doi.org/10.1080/07370020903586696

Tomlinson, B., Blevis, E., Nardi, B., Patterson, D. J., Silberman, M. S., & Pan, Y. (2013). Collapse informatics and practice: Theory, method, and design. *ACM Transactions on Computer-Human Interaction, 20*(4). http://doi.org/10.1145/2509404.2493431

Verbeek, P.-P. (2005). *What things do: Philosophical reflections on technology, agency, and design*. Pennsylvania State University Press.

Vyas, D. (2012). Domestic artefacts: Sustainability in the context of Indian middle class. In *Proceedings of the 4th International Conference on Intercultural Collaboration*. New York: ACM. http://doi.org/10.1145/2160881.2160900

Wakkary, R., Desjardins, A., Hauser, S., & Maestri, L. (2013). A sustainable design fiction: Green practices. *ACM Transactions on Computer-Human Interaction, 20*(4). http://doi.org/10.1145/2509404.2494265

Wakkary, R., & Maestri, L. (2007). The resourcefulness of everyday design. In *Proceedings of the 6th ACM SIGCHI Conference on Creativity & Cognition*. New York: ACM.

Wakkary, R., & Tanenbaum, K. (2009). A sustainable identity: The creativity of an everyday designer. In *Proceedings of the SIGCHI Conference on Human Factors in Computing Systems*. New York: ACM. http://doi.org/10.1145/1518701.1518761

Zhang, X., & Wakkary, R. (2011). Design analysis: Understanding e-waste recycling by Generation Y. *Proceedings of the Conference on Designing Pleasurable Products and Interfaces*, Article Number 6. http://doi.org/10.1145/2347504.2347511

Zimmerman, J., Forlizzi, J., & Evenson, S. (2007). Research through design as a method for interaction design research in HCI. In *Proceedings of the SIGCHI Conference on Human Factors in Computing Systems* (pp. 493–502). http://doi.org/10.1145/1240624.1240704

2 Towards a social practice theory perspective on sustainable HCI research and design

Adrian K. Clear and Rob Comber

This chapter grapples with a conundrum of scales in sustainable HCI (SHCI): on the one hand, it is now widely agreed that our current ways of living are unsustainable and that if we have any chance at a sustainable future, fundamental changes in these are required. On the other hand, many of the approaches that HCI research and design are taking in addressing sustainability challenges are much too limiting, in various ways, in the context of the scale of change required to avoid catastrophic climate change. In this chapter, we draw on social practice theory framings of energy use (and other resource consumption that leads to the emission of harmful greenhouse gases) to develop some assertions for what such theoretical underpinnings mean for the business of doing design for sustainability. We conclude that we need effective ways for doing transdisciplinary research and design; we need ways of working at broader temporal scales that account for historical and future trajectories of practice in design; and we propose a framework of social practice theory and design fiction as a promising approach to augment existing practice-oriented design.

Recently, a practice-oriented approach to sustainable HCI has begun to emerge (Pierce et al., 2013), which is part of a broader "practice turn" in HCI (Kuutti & Bannon, 2014) whereby researchers are moving to social science theories like social practice theory in order to explicate everyday life and its idiosyncratic complexities, to improve the products of design. This complexity of everyday life and the huge environmental challenges that society is facing are intertwined. It is estimated that to have a reasonable chance of avoiding catastrophic consequences of climate change – including extreme weather events, threats to our food and water supply, and threats to many plant and animal species – we can emit a further 565 gigatonnes of carbon dioxide – that is just 20% of the coal, oil and gas reserves of the world's fossil fuel companies; mitigating climate change disasters requires ways of living that that involve leaving most of our remaining fossil fuels in the ground (Berners-Lee & Clark, 2013).

Designing for new "ways of living" sits in contrast to the traditional "Interaction paradigm" in HCI in terms of the scope that it allows for change, by virtue of its framing of the design task and potential solution space: Kuutti and Bannon (2014) write that while the Interaction paradigm is about the application of technologies to change human actions, for the Practice paradigm, "*a whole practice is* the unit of intervention; not only technology, but everything related and interwoven in the performance is under scrutiny and potentially changeable, depending on the goals of the intervention." The Interaction paradigm in SHCI follows from an historical focus on cognitivism and information processing (Harrison et al., 2007). These approaches, themselves, are embedded in economic, political, and cultural systems, and they mirror the dominant paradigms of social change (i.e., economics and psychology) that are embedded in environmental policy (Shove, 2010). Their lure comes, in part, from promises of generalizability and large-scale behavior change that can be achieved by supporting and manipulating individual mental processes. The risk, of course, and this is one of the stickiest tension points for SHCI, is that everyday life is more complex than these models allow for and, hence, we ultimately get results and solutions that are inappropriate and ineffective. The Interaction paradigm isolates itself from any concerns with the cultural or the political, whereas these are integral parts of social practice and hence any paradigm that concerns itself with practices as analytical units (Kuutti & Bannon, 2014).

Practices might include things like commuting to work, bathing or making dinner, to give a few examples. Multiple theories of practice exist for formalizing these, as Pierce et al. (2013) allude to in their introduction to the journal special issue on "Practice-Oriented Approaches to Sustainable HCI". For the purposes of this chapter, we take Shove's framework that defines practices in terms of three elements: materials *"which "[include] things, technologies, tangible physical entities, and the stuff of which objects are made"*; competences, which *"encompass skill[s], know-how and technique[s]"*; and meanings, which include *"symbolic meanings, ideas and aspirations"* (Shove et al., 2012 in Pierce et al., 2013). Practices are socially constituted phenomena that characterize our everyday activities and routines. Hence, they are not the intention or action of an individual person but a collection of materials, competences, and meanings that come together and are reproduced in socially meaningful ways through the performance of activities. In terms of framing resource use, the main implication of a practice-oriented approach is that people do not use energy or resources – hence resource consumption cannot be reduced to individual choice; rather, resources are relied upon for the accomplishment of practices. And, agency in practices, and hence energy consumption, is distributed among the elements that make up the practice, which includes the material artifacts that are relied upon for its accomplishment (Hampton, 2017). For example, some established forms of bathing rely upon a shower and plumbing infrastructure in the home, water for washing and rinsing, and energy for heating water to an acceptable temperature. Each of these, in turn, has some agency in the reproduction of the practice. It is the social constitution of practices – that they are shared, normal, and routine ways of living – that bring large potential for scalable change for sustainability.

A practice-oriented approach, then, necessarily takes these entire practices as units of analysis, intervention and design. We see that it requires a broadening of our understanding of people and everyday life, from "users" and consumers, to account for the contexts in which resources are spent. Indeed, the importance of context in shaping our behaviors and interactions is not a new idea in HCI (Suchman, 1986). However, the context for sustainability is arguably unprecedented in the field of HCI. It is one of large-scale social change involving radically new forms of living (Shove, 2010; Kuijer et al., 2013) to address a "wicked" problem. Established methods for doing interaction design and achieving scalable results in traditional HCI contexts (e.g., laboratory studies for evaluating interface design for task completion, exercises inspired by specific domain understandings, or participatory and co-design with specific participant groups) are severely limited for challenges involving social change. In the rest of this chapter we ask, what does it actually mean to *take* a practice-oriented approach to HCI research and design for sustainability, and what can we hope to get out of it? We develop some assertions about this from literature on social practices.

A broader scope for design

Rather than being solely concerned with interactions between humans and computers, the concern of the HCI researcher and designer taking a practice approach becomes these wider configurations of practice elements, which include computers and people's interactions with them, but also a range of equally important elements including other forms or materiality, know-how, norms and expectations. As Pierce et al. (2013) put it, *"they expand beyond human-computer 'interactions' to grapple with the complexities of sustainability in terms of how people go about their everyday lives"*.

This represents a step-change in SHCI: in an effort to overcome the complexity of energy and climate change, a dominant approach in the field, following on from approaches in policy-oriented research (Shove, 2010), has been to narrow our conceptualizations of social change to one of individual behavior change. DiSalvo et al. (2010) found that in approximately 70% of SHCI literature, interventions were targeted at individual consumers. In this framing, the massive scale change required to mitigate harmful effects of climate change can be achieved through the aggregated effects of individual consumers making (technologically mediated) informed, rational decisions out of self-interest (Dourish, 2010). Interaction design in this vein might be categorized as "persuasive technologies" or "eco-feedback" (DiSalvo et al., 2010; Brynjarsdottir et al., 2012). The aim is to 'nudge' or bring about changes in individual consumer behavior through the provision of information about levels of resource consumption, using psychological and economic models of change. For example, an in-home display might provide householders with quantitative measures of current and historical electricity, gas, or water consumption.

The practice lens has been used to provide an account of everyday life and energy consumption that draws attention to the limitations of these particular intervention approaches, or framings of energy use, by highlighting the breadth of constraining forces that hold unsustainable practices together when the designer intends change (e.g., Clear et al., 2013; Strengers, 2014). Ultimately, what this approach reveals then, are understandings of the established, unwavering units of everyday life, dissected into the elements and their linkages that constitute practices. Our first assertion, then, is:

1   Social practices provide us with a framework for evaluating design interventions, by examining the ways that configurations of elements resist intended behavior change.

In this way, practices provide important domain understandings for *informing* research and design initiatives, and we must not undervalue this exercise, but our concern in this chapter is with what a practice-orientation tells us about *operationalizing* these in designing for sustainability. Matt Watson (2013) describes a practice as *"a concept which enables analytical attention to work on from specific moments and sites of action, to comprehend how moments and patterns of doing are orchestrated and reproduced over time and across different spaces"*. This description provides a coherent link between empirical inquiry and change at scale, and importantly, points to the magnitude and mechanics of the design task in question. The mechanism by which social practice opens up scalable change is in drawing out from the experienced reality or situated performance to extrapolate and intervene in the social fabric that holds ways of doing together, i.e., going from practice as performance (the physical act of doing) to redefine practices as entity (a shared notion of what constitutes a practice). For instance, observing cooking practices might reveal the use of particular ingredients, and characterize this use with the issues of its materiality (e.g., origin, availability, cost), competences (e.g., food literacy, cooking skill), and meanings (e.g., satiety, healthfulness, cultural appropriateness). These connections necessarily extend beyond the individual performance, for instance to infrastructures of food production and transport, economic and market configurations of access, and cultural values such as gendered roles which have implications for who and how we learn to cook. We see that:

2   Social practice theory provides designers with a framework to move from reasoning about design at the micro-level of experience and situated performance of practice, to the macro-level material and social fabric that holds 'ways of doing' together, and to understand interactions between these.

Important sites for engagement beyond individual consumers are obscured by narrow framings of consumption as individual choice. Brynjarsdottir et al. (2012) link underlying assumptions about rational choice, and dominant foci on calculability, efficiency, and top-down control, to modernism. Paul Dourish (2010) highlights how dominant cultures of market capitalism lead to intervention approaches that reduce social change for sustainability to the moral choice of individual consumers. By framing issues of climate change in terms of personal responsibility, such models ignore questions about the roles and responsibilities of other important social actors such as governments, institutions, and corporations, and, for example, their policies, regulations, and ideas of corporate responsibility in which our unsustainable ways of life are embedded and sustained. In fact, such framings serve in turn to sustain these forms of governance (Shove, 2010). And, included in this feedback loop are our notions about what valid approaches for modeling social change and designing interventions are. Before we can effectively engage with climate change issues, we need to be explicit about the extent to which our existing systems and social entities such as governments and corporations shape our everyday lives and patterns of consumption (Shove, 2010; Dourish, 2010).

Ultimately then, the task we are charged with is to redefine social life as we know it. Hence, as well as highlighting constraining factors outside of the control of the "users", a practice-orientation also alludes to the extent to which elements of everyday life and the ways that they are configured are beyond the control of the HCI researcher or designer. This represents a challenge for the SHCI community to find effective ways to do transdisciplinary design research.

3   Everyday life is shaped by our social systems and actors, meaning that much of the sustainable design task is out of the scope of any individual design practitioner or discipline.

In summary, practices are mostly comprised of elements that normally fit outside the scope of human-computer interaction, such as cultural reproduction of practice, infrastructures and policy, and the prevalence of neoliberal economics. We might question whether an exercise in sustainable change falls within the remit of HCI at all. This temptation highlights the transdisciplinarity of the challenge but also the need to acknowledge and be satisfied that while computers do not play a central role in the construction of our everyday lives, they, importantly, are relied upon, shape, and are shaped by the things that we do in direct and indirect ways. By corollary, their "design" shapes and is shaped by our (un)sustainable practices and ways of living. But "design" more broadly includes other material and nonmaterial elements and many other stakeholder groups (including the practitioners) are employed in doing the designing, and so we might consider practices to be the outcome or emergent behavior of the interactions between these processes. Our capacity and challenge as HCI researchers and designers is in understanding and manipulating interaction design within this broader context of designing sustainable living, and we might more effectively do so by coordinating stakeholders. The concerns, needs, values, and language of these may be diverse and conflicting, but we might leverage practices as an analytic entry- and endpoint of this broader system.

Considering temporality and change

If we take a social practice orientation to understanding the sustainability of resource consumption, we find that the elements of practices and their configurations have become established through historical trajectories and that any practice in time is linked in meaningful ways to its historical configurations. Take, for example, showering: Hand et al. (2005) provide an illuminating account of how current UK bathing practices, which involve one or two showers a day, have changed in less than a generation from a weekly bath. The prevalent material elements of this practice – en suite bathrooms and stand-alone power showers (we could perhaps now also include combi-boilers) – have emerged from bathtubs in shared bathrooms. During this emergence, bathtubs augmented with electric showers were popular. Conventions of short, convenient daily or twice-daily showers have emerged from lengthier, weekly baths. And, conventions of the body and hygiene have become more related to ideas of health and fitness.

Change in practices is always incremental (Watson, 2013). In thinking about HCI design for sustainability, this means that any future practices, and any digital artefacts and user experiences embedded in them, will have meaningful trajectories from current practices. Hence,

4   Design for future practices should be suitably contextualized and informed by understandings of current practices.

To elaborate on this point, the context of design that we refer to here might usefully be framed in terms of the inertia of the elements that constitute current practices. Hand et al. (2005) illustrate how practices have a momentum of their own in that they are continually reproduced through their performance in everyday life, giving them stability, or a 'closing' effect. They are also constituted by these performances, which can vary in more and less significant ways in any given instance. Repeated performances involving new elements can lead to fundamental changes in the practice, meaning that the future of any practice is 'open'. However, "'obdurate' elements have a 'closing' effect on the variety of plausible futures, so structuring otherwise 'open' possibilities" (Hand et al., 2005).

That changes in practice are incremental is not to say that design exercises should limit their scope to increments on the present, and the realms of plausibility for this. In fact, we argue for the contrary, as framings that treat too much of the present, such as current energy demands, as foregone conclusions only serve to reproduce it (Hand et al., 2005). We might argue that persuasive technologies are overly rooted in the present: while they can certainly be effective in particular situations where the targeted behavior is isolated and defined, in the context of sustainable living they limit the scope for reductions in resource use to easy and convenient shavings from the status quo. Any conservation changes will be made within individuals' own conceptions of need and what are normal ways of doing things – these are socially and culturally constructed, for example, in the case of thermal comfort (Shove, 2003), and it is these configurations of need and normality that are unsustainable. This approach treats the reduction of energy demand as an information deficit problem for "energy consumers" – that people will act as micro resource managers (or "Resource Man" (Strengers, 2014)) if the right information and tools are available to them. In reality, this atomization of behaviors, resources and actors takes a simplistic view of everyday life in which people can and will make easy decisions to optimize the energy consumption of their and their household's practices. Everyday life is more complex and socio-culturally varied (Brynjarsdottir et al., 2012) and so the causal relationship between information/feedback and what people do (behavior change) is often extremely weak – what people do is shaped by a complexity of factors including household dynamics, social and cultural conventions and values, routines and pressures on time. As such, interventions appeal to *"a limited number of people who are interested in their energy data and bills, who want to talk about their consumption with others, or who want to use new technologies to manage their energy demand"* (Strengers, 2014).

More fundamental changes in our ways of living than these approaches allow for are required. As Elizabeth Shove writes, for any effective response to climate change challenges, *"new forms of living, working, and playing will have to take hold across all sectors of society"* (2010). Silberman et al. (2014) argue that for HCI, this means *"considering, as rigorously as possible, the long-term social, economic, political, and ecological processes that might influence the adoption, use, and effects of particular technologies and practices"*. This suggests that for sustainability, we shouldn't limit ourselves by the present in design; we should consider long-term projections into the future, but social practice theory reminds us of the importance of considering *the process* of incremental change in practice to get there. In summary,

5   Sustainability requires radically new ways of living so design should not be limited (and limiting) by present configurations of practice and resource use, but it must consider the process of incremental change that stems from these to more sustainable ways of living.

According to Hand et al. (2005), the practice is not a consequence of any of its elements and their trajectories in isolation, but the coming together, the linkages, between these elements in time. Continuing with their bathing case study, prior to its current place in bathing practices, the shower already existed and featured centrally in bathing in Roman times, when bathing was public and collective, but then fell out of use. And the plumbing infrastructure required for showers existed long before their widespread adoption. Their reintroduction into common use occurred alongside wider changes in the temporal organization of daily life: "As such the shower belongs to a set of domestic devices whose popularity has grown precisely because they promise to help people cope with the temporal challenges of (late) modern life" (Warde, 1999 in Hand et al., 2005).

This implies that it is not sufficient to understand current practices for what they are now, as this tells us little about designing for change. For this, we need better understandings of how current practices have come to be, which includes how they are established through changes in historical practices. This requires not only a focus on the constituent elements and their respective historical trajectories. More importantly, it requires a focus on the linkages between the elements and how these are made and broken over time.

6   To design for a process of change towards more sustainable ways of living requires designing for the making and breaking of linkages between elements of practice.

Silberman et al. (2014) argue that "the processes that give rise to the issues indexed by the term 'sustainability' are larger in time, space, organizational scale, ontological diversity, and complexity than the scales and scopes addressed by traditional HCI design, evaluation, and fieldwork methods". Their conceptualization of time is, quite rightly, future-oriented, calling for research that considers longer timescales of change. However, if we consider how practices are established, this argument might also usefully extend to the past. First, for understanding how change happens, as we noted above. But also, many of our ways of living in even the recent past were more sustainable. There may be elements, or configurations of elements that were once part of acceptable ways of doing which we might aim to design back into future practices. Or we might gain valuable insight from tracing alternative trajectories of material artifacts that fit with the evolution of the other elements of a practice, much in the same vein as Steampunk culture preserves Victorian-era aesthetics in fictional futures (Tanenbaum et al., 2012).

7   Understandings of past practices can provide us with deeper roots for design exercises that aim to map possible future trajectories for sustainable practice.

We can sum up these assertions as tension points along two dimensions that arise from our pursuit for scalable change: abstraction (from concrete situated performances of practices to shared ideas about what the practice is), and temporality (from the very present situated performance to trajectories of change that project from the past into the future). First, the practice-as-entity represents practice-at-scale, as it is the social, shared understanding of a practice that enables us to recognize a practice performance for what it is, and continue to reproduce it – it is the situated performances of a practice that shape the practice entity. This implies that we must be analytically concerned with the situated performance, but that any innovation in performance will be limited by the forces that hold the elements of the practice together. And that while innovation/experimentation in any isolated performance might tell us something about plausible changes in a practice from the practitioner perspective, it excludes from design the more indirect social and cultural forces – and stakeholders in these – that shape the performance and the shared practice entity.

The second tension point concerns temporality. The need to be concerned with situated performance places a strong emphasis on the present. And, while we have established that situated performance and the present are of critical concern, such a focus can neglect longer-term change in designing for sustainability. In the next section, we address this challenge of maintaining a concern with the present while also integrating the longer-term future in the design exercise. And, of designing for future practices that are radically different, while maintaining sensible trajectories from current ones.

Innovation in practices

Thus far, we have motivated a concern with practices in HCI design for sustainability, and we have made some assertions about what the scope and dynamics involved in configuring practices mean for the design task. We have seen how notions of scale are problematized with the practice turn in illustrating how previous work that fits with the interaction paradigm has focused too narrowly. But, we also see how a broadening of this scope introduces new complexities of scale – elements, processes, and temporalities that are typically outside of the remit of HCI. In this section, we begin to address some of these in asking *how* to design in this context: what design approaches are appropriate for the innovation work involved in constructing and transitioning to new ways of doing.

The first tension point that we concluded the previous section with is recognised by Kuijer et al. (Kuijer & de Jong, 2009, 2011; Kuijer et al., 2013), who suggest a move from an analytic concern with practices to considering what it means to treat practices as units of design, "*generating and evaluating what could (or should) be in the future*". Their method for this relies on the bringing together of "*crises of routine*" and "*improvised performance*" of practice. A crisis of routine is required to "*overcome resistance to radical change*" and entails the orchestration of situations where practitioners can break out of "*existing material and social structures*". In their case study, Kuijer et al. (2013) used a laboratory environment to isolate the practitioner for this purpose. This setting consequently provides scope for improvised performance where the practitioner adapts and experiments with mutations of the practice, constrained by the lab environment and the structuring of the design task, to find the most valid and acceptable configurations for them. In their work, Kuijer et al. (2013) recruited improvisation actors to consider "splashing" as an alternative (to showering) bathing practice. We might draw parallels to a recent HCI study by Hasselqvist et al. (2016) where participants relinquished their cars and replaced these material elements of their transport practices with small electric vehicles for one year. In endeavoring for actual changes in practice, Kuijer et al. (2013) draw attention to the importance of 'doing' that is characteristic of these approaches. First, in that the designing becomes the concern of the practitioner and, secondly, the act of performance gets beyond people's imaginaries to what they actually do. In fact, what participants don't do might be as revealing about designing for sustainable practices as what they do.

This innovative approach to practice-oriented design embodies two important characteristics that relate to the assertions we derived previously. First, a concern with the 'doing', or the situated performance; and, second, a means (crisis of routine) to reduce the constraining impact of established practices on the innovation. Kuijer et al. (2013) remind us, however, of the gap that remains between innovation of this nature and the innovation of social change: "*Only if this variety on the practice is repeatedly performed by several practitioners (if it works and spreads), it can change from exceptional and improvisational to acceptable and normal, thereby reconfiguring the practice-as-entity.*" We can draw on the assertions we made previously to outline how this approach might be augmented to overcome some limiting factors in this.

We suggest that one promising approach to explore for this is to employ design fictions, written within a framework of social practice, as they have a number of characteristics that fit our purposes, here. First, they can serve as "boundary-negotiating objects" – "objects which both inhabit several intersecting social worlds and satisfy the informational requirements of each of them" (Star & Griesemer, 1989) – that can be co-created and negotiated by the various relevant stakeholders. In Kuijer et al. (2013), the innovation is restricted to the perspective of the practitioner. Design fictions enable the creation of crises of routine that are inclusive of the range of stakeholders that are invested in the configuration of the various elements that make up everyday practices. For example, for food practices, some of these might include domestic practitioners, vendors, marketers, farmers and nutritionists. Working successfully with a range of stakeholders to achieve purposeful improvisation requires common agendas that unite them, and climate change and sustainability, while being relevant backdrops for crises of routine, are often too distant from stakeholders' everyday concerns. Hence, one important role for design fictions is in constructing sustainability agendas and goals that are relevant for the range of stakeholder groups.

Design fictions also provide scope for imagining much larger-scale crises of routine, along the dimensions of abstraction and temporality that we mentioned previously. Related to the latter, they can extend well into the future while providing a narrative from the present, thus enabling us to treat sustainable design as a process. And, they could facilitate a means to interact with situated performance and vice versa, by providing a much richer framing of the kind of improvisational work in Kuijer et al.'s (2013) approach, while feeding the experience of this back into the shape of the narrative.

Design fictions can also help us design for the negotiation work involved in 'real world' improvisations of practice. Although crises of routine provide opportunities for reconsidering the status quo, in 'real world' innovation, it is impossible to separate the negotiation of acceptable and valid practice reconfigurations from existing social and material infrastructures. What we might strive to do instead is to draw attention to the wide variety of possible performances, and what factors are at play – not just the practicalities, but the politics involved – in prioritizing one over another. Design fictions can be diverse, considering a multiplicity of alternatives. They are diegetic prototypes in that they "explore alternate models of values and meanings" and these in turn cause us to reflect on and critique our own norms and values (Tanenbaum et al., 2012). Tanenbaum et al. (2012) show how the Steampunk community, through doing design fiction, come to establish their own cultural values through debate and negotiation about the cultural values and political meaning expressed, consciously or unconsciously in the designs that they create. In a similar vein, we might imagine how a community of stakeholders might establish their own sustainability values. For example, design fictions might allude to power dynamics, or inconsistencies between objectives and agendas within organizations. In this way, a key role for HCI might be a recursive one in inspiring crises for routine by bringing elements of the status quo into the spotlight.

In fact, we might look to our own communities of practice first. Sustainability is, at best, a fringe topic in HCI. What might a sustainability agenda look like for the ways in which we 'do' our work? Coming back to the quote that this chapter started with, might sustainability design fictions help us reconsider and establish new "patterns that perpetuate" themselves in our community that treat sustainability as a fundament? Of course, this would include fundamental questions about technology design, but it might also include questions about, for example, our publishing venues and cycles (Silberman et al., 2014).

Closing remarks

In 1950, Norbert Wiener characterized social life with the following quote: *"We are but whirlpools in a river of ever-flowing water . . . We are not stuff that abides, but patterns that perpetuate themselves"* (Capra, 1996). Ultimately, this chapter has highlighted the full complexity and huge challenge of what it means to consider HCI for breaking out of the seeming inertia of such patterns to bring about and perpetuate more sustainable ways of living. We have addressed the limitations of focuses on the individual and consumer, and offered an account of social practice as a pathway to considering change at a broader scale. Although the development of a practice-oriented approach for SHCI is a major progress in this direction, we are quite a way off in realizing the potential for sustainable design that the theory promises. In recognizing the transdisciplinarity of the challenge, it is clear that any route forward for research and design will entail meanderings of understandings and intervention around the complexity and depth of everyday practice elements and their interconnections. The widening lens of social practice means that we cannot always expect simple, easy, or quick solutions and research contributions will often not clearly fall into one discipline or another, or be expressible as quantifiable reductions in energy use or greenhouse gas emissions. However, social practice theory provides a framework that we can use to position and orient our approaches and contributions towards design for sustainable living, and as a means to link our new insights and understandings to the design of the digital materials that most concern us in HCI.

References

Berners-Lee, M., & Clark, D. (2013). *The burning question: We can't burn half the world's oil, coal and gas: So how do we quit?* Profile Books.

Brynjarsdottir, H., Håkansson, M., Pierce, J., Baumer, E., DiSalvo, C., & Sengers, P. (2012). Sustainably unpersuaded. In *Proceedings of the 2012 ACM Annual Conference on Human Factors in Computing Systems (CHI '12)* (p. 947). New York: ACM.

Capra, F. (1996). *The web of life: A new synthesis of mind and matter.* HarperCollins.

Clear, A. K., Hazas, M., Morley, J., Friday, A., & Bates, O. (2013, April). Domestic food and sustainable design: A study of university student cooking and its impacts. In *Proceedings of the SIGCHI Conference on Human Factors in Computing Systems* (pp. 2447–2456). ACM.

DiSalvo, C., Sengers, P., & Brynjarsdóttir, H. (2010, April). Mapping the landscape of sustainable HCI. In *Proceedings of the SIGCHI Conference on Human Factors in Computing Systems* (pp. 1975–1984). New York: ACM.

Dourish, P. (2010, August). HCI and environmental sustainability: The politics of design and the design of politics. In *Proceedings of the 8th ACM Conference on Designing Interactive Systems* (pp. 1–10). New York: ACM.

Gram-Hanssen, K. (2010). Standby consumption in households analyzed with a practice theory approach. *Journal of Industrial Ecology, 14*(1), 150–165.

Hampton, S. (2017). An ethnography of energy demand and working from home: Exploring the affective dimensions of social practice in the United Kingdom. *Energy Research & Social Science, 28*, 1–10.

Hand, M., Shove, E., & Southerton, D. (2005). Explaining showering: A discussion of the material, conventional, and temporal dimensions of practice. *Sociological Research Online, 10*(2).

Harrison, S., Tatar, D., & Sengers, P. (2007, April). The three paradigms of HCI. In *Alt. Chi. Session at the SIGCHI Conference on Human Factors in Computing Systems* (pp. 1–18). San Jose, CA.

Hasselqvist, H., Hesselgren, M., & Bogdan, C. (2016, May). Challenging the Car Norm: Opportunities for ICT to support sustainable transportation practices. In *Proceedings of the 2016 CHI Conference on Human Factors in Computing Systems* (pp. 1300–1311). New York: ACM.

Knowles, B., Blair, L., Coulton, P., & Lochrie, M. (2014, April). Rethinking plan A for sustainable HCI. In *Proceedings of the SIGCHI Conference on Human Factors in Computing Systems* (pp. 3593–3596). New York: ACM.

Kuijer, L., & De Jong, A. (2009, December). A practice oriented approach to user centered sustainable design. In *Proceedings of the 6th International Symposium on Environmentally Conscious Design and Inverse Manufacturing, 7–9 December 2009, Sapporo, Japan*. The Japan Society of Mechanical Engineers.

Kuijer, L., & De Jong, A. (2011). Exploring practices of thermal comfort for sustainable design. In *Workshop on Everyday Practice and Sustainable Design, CHI* (Vol. 201).

Kuijer, L., De Jong, A., & Eijk, D. V. (2013). Practices as a unit of design: An exploration of theoretical guidelines in a study on bathing. *ACM Transactions on Computer-Human Interaction (TOCHI), 20*(4), 21.

Kuutti, K., & Bannon, L. J. (2014, April). The turn to practice in HCI: Towards a research agenda. In *Proceedings of the 32nd Annual ACM Conference on Human Factors in Computing Systems* (pp. 3543–3552). New York: ACM.

Morley, J., & Hazas, M. (2011). *The significance of difference: Understanding variation in household energy consumption*. ECEEE.

Pierce, J., Strengers, Y., Sengers, P., & Bødker, S. (2013). Introduction to the special issue on practice-oriented approaches to sustainable HCI. *ACM Transactions on Computer-Human Interaction (TOCHI), 20*(4), 20.

Shove, E. (2003). *Comfort, cleanliness and convenience: The social organization of normality* (Vol. 810). Oxford: Berg.

Shove, E. (2010). Beyond the ABC: Climate change policy and theories of social change. *Environment and planning A, 42*(6), 1273–1285.

Shove, E., Pantzar, M., & Watson, M. (2012). *The dynamics of social practice*. Sage.

Silberman, M., Nathan, L., Knowles, B., Bendor, R., Clear, A., Håkansson, M., ... Mankoff, J. (2014). Next steps for sustainable HCI. *Interactions, 21*(5), 66–69.

Star, S. L., & Griesemer, J. R. (1989). Institutional ecology, translations and boundary objects: Amateurs and professionals in Berkeley's Museum of Vertebrate Zoology, 1907–39. *Social Studies of Science, 19*(3), 387–420.

Strengers, Y. A. (2011, May). Designing eco-feedback systems for everyday life. In *Proceedings of the SIGCHI Conference on Human Factors in Computing Systems* (pp. 2135–2144). New York: ACM.

Strengers, Y. A. (2014). Smart energy in everyday life: Are you designing for resource man? *Interactions, 21*(4), 24–31.

Suchman, L. (1986). *Plans and situated actions*. New York: Cambridge University Press.

Tanenbaum, J., Tanenbaum, K., & Wakkary, R. (2012, May). Steampunk as design fiction. In *Proceedings of the SIGCHI Conference on Human Factors in Computing Systems* (pp. 1583–1592). New York: ACM.

Watson, M. (2013). Commentary III: Theories of practice, everyday life and design futures. *ACM Transactions on Computer-Human Interaction (TOCHI), 20*(4), 20.

3 A conversation between two sustainable HCI researchers

The role of HCI in a positive socio-ecological transformation

Samuel Mann and Oliver Bates

ACT I: Introduction

Oliver: Have you seen Mike Hazas and Lisa Nathan's call for papers?

Sam: Yes, I love the premise.

Oliver: It's what we've been talking about. Do we write a chapter repositioning human-computer interaction (HCI) as an activist discipline?

Sam: The call for papers starts with a promise, "Digital technologies are hailed as revolutionary solutions to problems of environmental sustainability. Yet" (M. Hazas, personal communication, April 6, 2016). And that big 'yet' is the problem. Mike and Lisa are asking us to do something that's different. They are asking us to do something that is raising conundrums – is that the words they used? Raising the paradox of conflict and being disruptive.

Oliver: Do you think that HCI hasn't been disruptive enough?

Sam: All right. If you look at the areas that they are asking for in terms of critical, ethical reflections, the political, shifting orientation, shifting the norms, the role of activism, do we think that HCI has been able to deliver at scale on any of those things so far?

Oliver: No. I don't think so.

Sam: Me neither, so our contribution wouldn't be on any of those themes individually but all of them together. Why haven't we been able have impact? Do we write a chapter repositioning HCI as an activist discipline?

Oliver: I remember you interviewing me two years ago and asking me if I was an activist. I said no. I thought I was just doing this because I was interested, not because I was trying to change anything. The more conversations I have with people and the more I got fired up about it, I realized activism is about actively seeking to change things.

Sam: Actively seeking to change things is a personal statement. Maybe that is that strong statement our chapter could make, that this is personal, who are we and why do we care? Then there's seeking to change things. What is it that we are seeking to change?

Figure 3.1 One CHI-soaked evening

Oliver: That kind of complexity around the day-to-day use of these technologies, and why it continues to spread and grow, how it has become ubiquitous fascinates me. I started out looking at how people used energy in the home, and classifying that in terms of social practices or services to try and understand how daily energy demand was constructed. Then I got more interested in people, the way they reflected on their use of digital technology and day-to-day life. I focused on interactions with digital technology and how that is a complex ecosystem (Bates et al., 2014). There are some seriously energy-intensive components of digital technology and its use, the manufacture, the data side of things. There are some great reductions that it can encourage too!

My strengths are in data, and understanding the complexity.

I want to do the right thing. First, what is the right thing? That's what makes the sustainability agenda in computing so hard. We want innovation to provide solutions but we also see innovation as a cause of lot of

Sam: the environmental problems. What I would like to do is to start with that which I am increasingly disillusioned with – computing's ability to deliver the "stuff".

Sam: Then we should talk about briefly about the promise of computing. And why hasn't it delivered? And why, despite this increasing disillusionment with computing's ability to deliver on this promise, we want to take an optimistic course.

Oliver: Not just that, there's a second message – the value of conversation.

Sam: Yes. We need to be talking with each other and colleagues from other disciplines about HCI's role in a sustainable future.

Oliver: Like you do in your conversations for Sustainable Lens. You have, what, 300 podcasted conversations with people from architects to zoologists?

Sam: And everything in between. OK, so wherever possible we'll link to conversations we've had. So, it'll be a conversation-based optimistic take on computing's possible contribution to a positive future...

ACT II: The promise (and problems) of computing

Oliver: Is it computing or is it HCI?

Sam: Well, if we define HCI as the relationship between computers and society rather than purely interaction design, or...

Oliver: Society is important! I think it's the clue to everything we are trying to do, empowering society to do stuff.

Sam: Then I'd like to talk about a definition of sustainability being ethics across time and space, and use that as a way into how it's about positive change at scale. How can we get there? And that's into the policy-influencing decision tipping point, focusing on things that are going to get us into those tipping points and tip them.

Oliver: Within that there's work that's not necessarily straight technology development but more along the lines of what people are thinking about.

Sam: It's work like Allison Druin's work with children (2016), Batya Friedman's work in both value sensitive design, and multi-lifespan information systems (2016) and Birgit Penzenstadler's software engineering work on getting sustainability into system requirements (2016). These are the things that we need to be getting in bulk.

Oliver: I think HCI is well suited for that. There were many people doing sustainability research, but at the last conference there were only a few papers. It's frustrating. Several authors have suggested directions that our research should be moving in, but it feels like a growing divide. Maybe more activism and being radical are a direction I should consider, but I'm concerned that if I pivot away from my current career trajectory that I'll compromise myself in some way. I don't want to pivot away from sustainable HCI, but to keep my career on course I think I must.

Sam: Wait, you can't pivot away now! I thought I was leaving it in safe hands.

Oliver: What do you mean, leaving it? But you wrote the first papers on computing education for sustainability (Mann, Smith, & Muller, 2008), you defined the sustainable practitioner and challenged computing to see themselves as sustainable practitioners (Mann, 2011). Are you going back to computing? Joining the crowd that only does sustainability as a hobby alongside their real jobs?

Sam: No, the other way. I'm not leaving sustainability. It's computing I'm having doubts about (Mann, 2016). I was only ever in computing to make a difference. I'm doubting our ability to drive change. My work is on the notion of the sustainable practitioner or the sustainable lens being the way that you do that. You can think of all of education being about the development of your lens, how you see the world, your values, your mindset. The early bits of life and education are about developing your personal lens, mixing your values with opportunities to practice increasingly advanced sets of skills. Then higher education is developing professional lens. The focus of my work is about making sure that that personal and professional lens is also a sustainable lens. For me the most important thing I can be doing is identifying and helping the disciplines that have got the greatest societal leverage. That's the disciplines where the handprint, the potential to do good is massively greater than the negative impact, the footprint. I think that education, and computing, in particular HCI, are disciplines with very high societal leverage.

Oliver: Yes. Is this how we describe the promise of computing, having high leverage with a large handprint?

Sam: For the moment, let's just look at anthropogenic greenhouse gas emissions. Barath Raghavan (2016) and Chris Preist (2013) have come up with a similar numbers, that the global impact of computing, is it about 2% of global emissions?

Oliver: That's nothing.

Sam: Two percent, that's hardly anything but it's the same as the airline industry. We are desperately worried about air miles and so on, and computing is as bad.

Oliver: It looks like it's growing too, I mean in terms of core infrastructure, the number of connected devices we have in the house, tablets, and everything.

Sam: We're struggling with computing's footprint reduction but the much bigger potential impact is computing's ability to do good, computing's potential to reduce total carbon emissions – people are talking about 16 to 40%. That's massive. Skip Laitner estimates that our society runs on about 14% energy efficiency; he says, "We are wasting most of what we produce" (Laitner, 2014).

Oliver: Two percent through footprint or more than 10 times that through handprint.

Sam: Most sustainable computing was about reducing the footprint of computing itself – remember all the fuss about virtualization?

Oliver: And people were doing behavior change work, first on the idea that people didn't know, so it was information, then using persuasion. "Sustainably Unpersuaded" (Brynjarsdottir et al., 2012) presented a critical view, questioning whether persuasion and behavior change had any real effect on the global problems of sustainability. The recommendations of their work have widely been taken on board, with the most obvious example being those who use participatory design. We're framing research around practices instead of behaviors; developing implications of the dynamics of practices changing over time; and, increasingly authors who reflect on their own studies and findings.

Sam: Is this a good thing?

Oliver: Well yes, and no. Yes, as people have broadened understanding of persuasion, including users in the design process. So, participatory design, co-creating, co-design is something that HCI does. Moving beyond the individual, that's something we struggle with. It's hard to do things of scale and represent demographics in the correct proportions, but then they also suggest that we should shift from prescription to reflection. Then there's the 'No'. There's been a dramatic reduction in SHCI papers since that paper. Were they perhaps too critical of an emergent research focus?

Sam: I think that we've gotten ourselves stuck on efficiency, about computing's potential positive impact, and somehow got into our heads that apps that tell you how long you've been in the shower are going to save the world. If it was going to save the world, how come the world's not saved yet? I've been thinking about this, and I think that worse than being ineffective, the focus on computing-supported efficiencies are doing us harm. There are several reasons.

First, the values work. Bran Knowles' work (2014), based on the work of people such as Pella Thiel (2015). The rational, economic man approach appealing to people's wallet is disabling the altruistic "we need to be doing this because it's what we need to be doing justification". HCI hasn't moved beyond that selfish individualistic approach.

Second, I think that the notion that HCI computing can deliver all of these efficiency gains through behavior change, or however they are going to achieve that, is what Susan Krumdieck refers to as a green myth (2015). It's the miracle just around the corner.

Oliver: We are all going to have electric cars one day so let's keep building roads.

Sam: Yes. There's this notion that computing is going to save us. Kentaro Toyama talks about the "geek heresy" (2016) – we think throwing technology at problems is going to solve it but his summary is technology exemplifies underlying human forces. If we are continuing to consume then throwing technology at it is not going to solve that problem.

Third, there's the substitution effects that we see all over the place.

Oliver: We found in both the domestic and non-domestic spaces there's a tension between installing the 'state-of-the-art' IoT infrastructure and using the minimum set of monitoring and control infrastructure to make better decisions around buildings and energy (Bates & Friday, 2017).

Sam: Yes, individual devices might be more efficient but then we massively increase the bandwidth around it. We are not actually considering it in terms of absolute terms, in terms of planetary boundaries.

Oliver: It's the area under the curve, right. That's how we frame our work by reminding the readers that people look at the instantaneous savings but the area under the curve is massive because on aggregate, you're doing so many more things that consume.

Sam: Which leads to number four: negativity. As Bob Costanza argues, sustainability is positive (2016).

Oliver: How's your list going? Sam's list of what's wrong with efficiency-focused HCI. I've got down rational resource man, green myth, substitution, negativity. What's five in the great list of what we're doing wrong?

Sam: Five. Perhaps the biggest issue of all is the very nature of IT. All of HCI is absolutely done in a context of sell more stuff.

Oliver: On the fringe, there is the sort of un-designed movement (Pierce, 2012) or the related slow design, but it's certainly not gotten major traction. We used those ideas as a framework to think about how to design technology, and for considering when designing something new is not necessarily the right thing to do.

Sam: Then there's the rebound effects like Airbnb causing more flights because it's easier to stay cheaply.

Oliver: I think it's the kind of one-size-fits-all kind of way of thinking. I could build an app that's community-driven and empowers people and society to make the 'right decisions', but are you empowering people, or are you empowering capitalism?

Sam: That brings up some of the nature of sustainability problems, which by definition aren't amenable to the paradigm that we generally work under: the clearly identify a problem, design a product, do some sort of experiment to measure an intervention.

Oliver: Yes. If the rebound is six, the nature of sustainability makes seven.

Sam: I'm excited by the work that focuses on community engagement, not as a means for behavior change, but for the sake of an empowered community. Steve Benford's trajectories and uncomfortable interactions (2013), Lancaster's Tiree work I talked to Maria Ferrario (2015) about.

Oliver: Yes, plus Rob Comber empowering communities (2016) and Dave Green (2016) – these are all good things, supporting community, beyond an experimental paradigm.

Sam: So those things are in response to that problem of the nature of sustainability. I think that the last problem, the eighth is that HCI in general has a weak understanding of sustainability. Weak as a technical term, as opposed to Daly's strong sustainability, the bullseye model, but also meaning limited understanding of sustainability. HCI seems to have gotten totally distracted by carbon, framing sustainability as only carbon efficiency for climate change. Yes important but framing it that way also ignores all the other important effects. How can HCI help biodiversity loss, global inequities?

Oliver: Social change?

Sam: Intergenerational equity. Perhaps that's what sustainability really means and yet I don't know that anybody with a possible exception of Batya Friedman (2016) and Lisa Nathan's (2012) multi-lifespan information systems that have really looked at how we might start to address that. They identify three categories of problem that we're unlikely to solve in a single human lifespan: (1) limitations of the human psyche . . . lasting peace . . . the first generation agree to keeping children alive, the second generation grow up in environment where they don't feel threatened, maybe third generation can really build a peace; (2) tears in social fabric; and (3) environmental timescales.

Oliver: It feels like an uphill battle to leave the world in a better state than when you joined it, or to be better for future generations.

Sam: Sustainability is, of course, the attempt to solve that wicked problem. It's important that we cast sustainability as the positive solution.

Oliver: I agree.

Sam: That's the nature of those wicked problems, the intergenerational timescales, the complex systems. While we have an approach to HCI which is focusing primarily on things that we know we can fix, we're not going to get there.

We've managed to depress ourselves. Where can we look for solutions for how HCI could better approach this?

Oliver: In 2014, after the workshop at CHI, Six Silberman set about outlining a set of 'next steps' for sustainable HCI (Silberman et al., 2014). The paper attempts to rally those who work in sustainable HCI in order to bring a critical mass back to sustainable HCI and sustainable interaction design.

Sam: Yes, I talked with Six about how problems of sustainability are larger in time, space, organizational scale, diversity, and complexity than the scales and scopes addressed by traditional HCI design, evaluation, and fieldwork methods (Silberman, 2012). That I would have thought was a big enough intellectual challenge to spur on the field.

Oliver: Yes and under those wider understandings of sustainability, there are also a large number of other societal issues that link to problems of sustainability (e.g., equality and race, feminism and gender, food, land, water, ICT4Development), and with those issues there is a number of HCI researchers who are already doing 'good' and helping change be affected in these areas. Do we still think we can get HCI together around the three pillars approach to sustainability?

Sam: Don't get me started on the nonsense that is a pillars approach to "environmental sustainability". You heard me cheer when Hans Bruyninckx told a conference last year that "a pillars model is intellectual nonsense, you cannot have a little bit of sustainability on a finite planet" (Bruyninckx, 2015).

Oliver: It would be remiss of us not to mention Eli Blevis' rubric. In 2007 he placed sustainability as a core principle for HCI design and research, not just an afterthought. This year Chris Preist (2013) and Daniel Schien (2016), working with Eli, added additional rubrics and guidelines that emphasize more responsibility around the environmental impacts of the digital technology, data demand, and reliance upon digital services. Perhaps the most understated and implicit takeaway from these two papers is that everything that we do or make has an environmental impact.

Sam: But it is still largely concerned with reducing computing's own footprint. Ecological economist Sigrid Stagl (2007) wrote a piece in an education journal that has always stuck with me. It's an unpacking of the challenge of impoverished visions of sustainability that Eric Myers and Lisa Nathan raise (2016). It states that as a society we must learn to live in a complex world of interdependent systems with high uncertainties and multiple legitimate interests. We should think simultaneously of drivers and impacts of our actions across scales and barriers of space, time, culture, species and disciplinary boundaries. It means that we need to switch from a focus on outcomes to one of process.

Oliver: That's far away from an app that tells you to get out of the shower. But it's quite long. I'll never remember that.

Sam: OK, how about system conditions? Karl-Henrik Robèrt (2015) created the Framework for Strategic Sustainable Development. He told me we need to backcast from a sustainable world on the basis of living within system conditions. Göran Broman (2014) describes it with the metaphor of chess: checkmate is defined in a principled way, as system conditions. Winning chess, or reaching a sustainable society without any idea of the principles that define that situation would be very unlikely. So, let's see HCI adopt the system conditions.

Oliver: Still pretty long.

Sam: OK, shorter, and my current favorite definition of sustainability, which I've borrowed and adapted from Albert Norström (2015) is that sustainability is about a positive socio-ecological transformation.

Oliver: Is that our chapter title? Our contribution? The role of HCI in a positive socio-ecological transformation? Sustainability as innovation?

Sam: I like it. It says it is about change, but that is positive. While it's a serious predicament that needs major a change, we're not ready to give up. Perhaps we're a balance to Collapse Informatics that Bill Tomlinson (2012) promotes.

Oliver: Some would argue that Collapse Informatics is trying to help positivity in times of collapse. I think it's great for a headline, kind of like, Doomsday. The world is on fire, we need to limit everything.

Sam: Peter Garrett didn't just sing, "How do we sleep while our beds are burning?" then build a bolt-hole in the desert. He campaigned and got himself into government. I think that what we need to be focusing on are the sorts of movements like the Transition Town movement. They're definitely not about setting up some sort of post-apocalyptic society. They're also not about convincing anybody that they have to change behavior. They're about saying, "We're happily demonstrating a better way of living."

Oliver: *The time has come*
To say fair's fair
To pay the rent
To pay our share

Sam: We're trying to be disruptive, not write a write a musical. One of the issues is the flip side of that more technology is the future and computing is all about that, is that, the notion that you're talking about sustainability, you want me to start living in a cave again. If we're going to find a role for HCI in a sustainable future, we need to find a solution to the double paradox of sustainability being about going backwards (but not going backwards) with sustainable HCI being about new ideas (but not new stuff).

Oliver: You jumped about doing a happy dance when I said, "sustainability as innovation" before.

Sam: I love that. I think what we need to be looking for positive precedents. We need to be looking for examples of people that are changing systems in positive values–based way. Perhaps if everyone in HCI could make a point of seeking out people working and living towards a transition. I visited Transition Town Oamaru and was blown away by the energy and passion and people such as Gail May-Sherman (2015) making a difference.

Oliver: That's OK for a town that has community values, but what about a business or other similar institutions?

Sam: There's Wishbone Design Studio (Latham & McIvor, 2016) a sustainable bike company. Their dream was for a product that would last from ages one to five, and then be passed on to the next young rider, a 100% repairable product that would never end up in the landfill and they actively promote a second-hand market. The role of sustainability is also about values in the operation of the business and the relationship with customers.

Oliver: Why do we need to keep perpetuating growth when actually it's about building a system where it's kind of closed and innovating within that?

Sam: Or the Interface Carpet story of everlasting carpet as a service. In computing, we're only just seeing a slight nod to that with device loyalty that Christian Remy (2016) describes. And not at all in software, not so much the software itself, but driving a positive system change. The question is, how do we get this positive system change, at scale, via HCI?

Oliver: Is HCI the vehicle for that? What would it look like for HCI to be run on the same basis as that town, or that bike shop?

Sam: Well, back to what Kentaro Toyama said, that "technology exemplifies underlying human forces" (2016). Technology can make a difference if it works with those human forces, not trying to superimpose itself on top of them.

Oliver: We need a coherent kind of mission statement. Better branding and direction. It's not clear, as a community, where we are, what we do, what we all want to do.

ACT III: Transformation

Sam: If we take as a starting point that sustainability is about a positive socio-ecological transformation.

Oliver: Yes.

Sam: It is about change. Yes, economics can be in there but only to achieving socio-ecological transformation.

Oliver: Where do the values fit in?

Sam: Phil Osborne talks about the relationship between values and value (Osborne, 2015). He says the production view of marketing has flipped to the service-dominant logic of exchange.

Oliver: How does this help sustainable HCI?

Sam: We need to see HCI as a service. Not about products, but in the service of sustainable transformation. How about a criterion for all HCI, how does this contribute to that positive socio-ecological transformation?

Did you notice that I didn't say sustainable HCI? Because I think that the key is that we need to normalize this. This isn't just people working on sustainability; this should be applied to all HCI research.

Oliver: I could apply to all digital technology or technology in general, HCI as the link.

Sam: How can we get away with publishing papers that are blatantly unsustainable? I think if we could convince people to put that in as a criterion we would make great strides. Perhaps a positioning in terms of Bob Willard's sustainability and maturity matrix (2005), going from avoidance, compliance, efficiency, opportunity, the reason for doing. At least it would be able to see to what extent is this contributing. Like the energy stars on the fridge, you'd be able to make informed choices about the value of the work. And like the whiteware manufacturers, this would drive innovation.

Oliver: If that's your measure stick, how will people measure their walk against it?

Sam: We would give them a rubric. Almost every paper that does consider sustainability and HCI starts with a definition of sustainability, generally a pretty weak one. It might change if we were to provide people

	with a way of saying, "This paper sits as a 3.5 on the Bates and Mann sustainable HCI maturity scale."
Sam:	If we base it on Willard's five-stage maturity model for sustainable business, the first one is, Stage 1: The company feels no obligation beyond profits. It ignores sustainability and actively fights against related regulations.

So that becomes:

Stage 1:	The researcher feels no obligation beyond publication. Research focuses on development of products without regard to wider implications. Ignores sustainability and ethics and/or actively argues that it doesn't apply this research.
Oliver:	OK, I got this, the second stage: The researcher manages their ethical responsibilities as compliance. The researchers are aware of implications and perhaps include a small section in the discussion that acknowledges a single sustainability factor, but wasn't incorporated into the research question, methods or outcomes.
Sam:	The third is the shower app.
Oliver:	Stage 3: Research is about products that deliver incremental, continuous improvements in eco-efficiency. Sustainability as an opportunity to explore aspects of HCI: encouraging behavior change or different ways to communicate. Little attempt to question whether the activity being made efficient is sustainable in wider terms, nor alternative approaches. If sustainability is defined, then it's pillars or "environmental sustainability".
Sam:	Research in Stage 4 is research on sustainability, using HCI. It references complex sustainability themes. It is value-driven. It adopts holistic sustainability, integrating all aspects through a process-based approach. In this stage, different models become apparent, understanding products as a means to deliver a service to a customer. Researchers are more likely to talk about empowerment, democracy, participation, and social systems than they are interventions for behavior change.
Oliver:	That's quite a leap from Stage 3.
Sam:	Yes, Willard describes that as a transformation. The early steps were transitions. Moving from Stage 3 to Stage 4 requires internalizing sustainability notions in profound ways, both personally and organizationally. The transformation to Stage 5 is one of the positive socio-ecological transformation being the premise of the research. Driven by a passionate, values-based commitment to improving the well-being of society, and the environment, the research helps build a better world because it is the right thing to do.
Oliver:	So, Stage 4 researchers do the right things so that they are successful researchers. Stage 5 researchers are successful researchers so that they can continue to 'do the right things'.
Sam:	Yes, where the right things are to drive a positive socio-ecological transformation.

ACT IV: Models and metrics

Oliver: I think we've got ourselves a rubric! You know how you must declare keywords at the start of the paper, you know classifications? Can the Mann and Bates thing be just there, you know?

Sam: There was a paper submitted to the sustainability track last year; it was on an App that helped you find a car park. I struggle to see how that had much to do with sustainability. The Mann and Bates metric might put that as a deluded 2 at best.

Oliver: Is there another dimension, something about scale of impact?

Sam: If we get back to that positive system change at scale, there are different ways you can achieve that. Remember how I talked to the head of the European Environment Agency, Hans Bruyninckx (2015)?

He said what they need to do is influence the 28 decision-makers in European Environment Ministries. How could HCI help that happen? Robert Brewer (2014) has doggedly asked that at the last couple of conferences, how can we have impact at scale?

Oliver: That is the question, right. I guess do we get more involved in politics. Are we saying HCI needs to struggle with that?

Sam: We need to find a way in. Whatever information systems that politicians are using to decide a particular vote, let's make sure that it promotes joined up, integrated thinking and includes a sustainability metric.

Oliver: You've been writing letters to the editor again.

Sam: Yes, but in terms of impact, imagine if the default for systems such as that supporting international trade could be. You could do it that way, but it would be more sustainable to do it this way. I talked to Dan Russell (2016), who works on improving search quality for Google. We called our talk "Searching for Sustainability". I asked him what would Google tell you if you asked, "Should I ride my bike to work today?" At the moment it will interpret it as a weather question. Imagine if it said, "Yeah, you should, otherwise your footprint this week will overshoot by 3000."

Oliver: That would be an interesting challenge.

Sam: Exactly. A transformative HCI is much more interesting than the shower app. You know when you do a search on Google for a movie, it tells you as a fact on the home page. If you do a search on, is climate change real? why doesn't Google come up with "Yes"? The potential is for influence at enormous scale. How can we help Google move to that position where they decide that that's the thing to do? That would have a massive impact.

Oliver: Isn't there something terrifying about large corporations tailoring answers and information.

Sam: Again, it's a messy question of information that is at the heart of sustainability. In *The Virtues of Ignorance* Bill Vitek and Wes Jackson (2008) argue that a "knowledge-based worldview is both flawed and dangerous". They argued for ignorance-based worldview: what would human cultures look like, if we began every endeavor and conversation with

	the humbling assumption that human understanding is limited by an ignorance that no amount of additional information can mitigate?
Oliver:	HCI has dabbled in that; we've seen some design fiction work recently. But is ignorance defeatist?
Sam:	Yeah, I think the book should have been called *The Virtues of Humility*. But the book argued ignorance celebrates knowledge (Vitek, 2008).
Oliver:	That frames what most of HCI is doing, not testing, not even stating "invent more widgets" as a paradigm.
Sam:	Paul Heltne (2008) says that ignorance is humble: "Acknowledging that one does not know is a humble kind of ignorance, one that is, in fact, filled often with the joy of discovery and wonder at what is discovered . . . This is the kind of ignorance-based worldview that can help us fathom the messes we are in, articulate assumptions and processes, entertain questions and be enriched by them, and imagine new ways and new knowledge" (Heltne, 2008).
Oliver:	This is what we need to inspire HCI.
Sam:	Robert Root-Bernstein (2008) argues that science is not a search for solutions but a search for answerable questions – it must become acceptable to say "I don't know". He says, "[S]cience is a way of asking more and more meaningful questions."
Oliver:	The same challenge goes for HCI research. Can we just replace science and education with HCI? HCI is about more meaningful questions and developing skills for critical evaluation. I think we can, but what are we going to do about it?
Sam:	Well, we're going to write this chapter that discusses where we think we're going. Alison Druin is working with the US National Parks Service, reworking their digital strategy. She says at first people said, "Well what you should do is just start small. Do something small; have a success of it", but her response is "But the very issue is that there are already pockets of good stuff but there isn't a system of good stuff" (Druin, 2016). We can expand what Alison said to wider HCI; we could cherry-pick things that are working. But there's an awful lot of areas where there's not good stuff happening. We need to focus on system change.
Oliver:	As you say, cherry-picking leaves too many gaps. Look at it in an HCI context: You get six sustainable HCI papers and then 300 talking about producing more of widgets without any consideration of environmental impact. And those six sustainable HCI papers will almost certainly be about energy efficiency and not considering the wider, deeper, transformative sustainability we've been talking about.
Sam:	One of the key things we've talked about is that relationship between sustainability and innovation, looking for positive alternatives. Transition Towns, Wishbone.

What would it take to create a positive system change at scale? |
| *Oliver:* | For a concrete contribution for HCI we talked about the Mann and Bates scale of sustainability. If you want to move a community towards something, you must start somewhere, right? |

A conversation between two sustainable HCI researchers 57

Sam: So if we had our rubric and we could ask people to somehow represent which of the UN's Sustainable Development Goals they're addressing through their work. They might say, "Well, actually it is energy, and it's not all these other things." Just doing that would be a useful thing for people to do.

Oliver: I agree with you. We are being active. We hosted a workshop on generating patterns around sustainability in HCI. Also, our #shcipat hashtag, where we tweeted proto-patterns from every talk we attended. The response showed there is interest and energy around broader conceptualizations of sustainability. It helped spark several conversations with presenters who've never even considered sustainable HCI.

ACT V: Conclusion

Sam: We started out saying it's personal, that we could be activists.

Oliver: How can we be more activist-y? We could rank every paper in the conference against our maturity rubric. Go around and put the appropriate

Figure 3.2 Mann-Bates sustainable HCI rubric

	number of stickers on the door while they're talking. That might inspire some people to explore attitudes towards sustainability. We'd wind some people up. At least it would start a discussion around it. Talking of winding people up, we should write that chapter – it would be awesome.
Sam:	What would be awesome would be a chapter that explored these ideas as a conversation, reflecting on the challenges, raising questions, not presenting fixed, perfectly formed answers.
Oliver:	Absolutely. We could finish with the rubric and then what Donald Norman said on your show: "Let's be good together instead of narrow minded and apart" (2014).

References

Bates, O., & Friday, A. (2017). Beyond data in the Smart City: Repurposing existing campus IoT. *IEEE Pervasive Computing, 16*(2), 54–60.

Bates, O., Hazas, M., Friday, A., Morley, J., & Clear, A. K. (2014). Towards an holistic view of the energy and environmental impacts of domestic media and IT. In *Proceedings of the 32nd annual ACM Conference on Human Factors in Computing Systems* (pp. 1173–1182). New York: ACM.

Benford, S. (2013, July 5). Experiencing changing trajectories. *Sustainable Lens* [Podcast]. Retrieved from http://sustainablelens.org/?p=505

Blevis, E. (2007). Sustainable interaction design: Invention & disposal, renewal & reuse. In *Proceedings of the SIGCHI Conference on Human Factors in Computing Systems* (pp. 503–512). New York: ACM. https://doi.org/10.1145/1240624.1240705

Brewer, R. (2014, June 3). Energy literacy. *Sustainable Lens* [Podcast]. Retrieved from http://sustainablelens.org/?p=669

Broman, G. (2014, December 20). Strategic sustainable development. *Sustainable Lens* [Podcast]. Retrieved from http://sustainablelens.org/?p=767

Bruyninckx, H. (2015, September 9). Challenging deep assumptions. *Sustainable Lens* [Podcast]. Retrieved from http://sustainablelens.org/?p=914

Brynjarsdottir, H., Håkansson, M., Pierce, J., Baumer, E., DiSalvo, C., & Sengers, P. (2012, May). Sustainably unpersuaded: How persuasion narrows our vision of sustainability. In *Proceedings of the SIGCHI Conference on Human Factors in Computing Systems* (pp. 947–956). New York: ACM.

Comber, R. (2016, March 19). Empowering communities. *Sustainable Lens* [Podcast]. Retrieved from http://sustainablelens.org/?p=1000

Constanza, R. (2016, July 29). Positive systems thinking. *Sustainable Lens* [Podcast]. Retrieved from http://sustainablelens.org/?p=1074

Druin, A. (2016, June 26). Children as design partners in technology and sustainability. *Sustainable Lens* [Podcast]. Retrieved from http://sustainablelens.org/?p=1016

Ferrario, M. A. (2015, February 15). Participating co-developers. *Sustainable Lens* [Podcast]. Retrieved from http://sustainablelens.org/?p=808

Friedman, B. (2016, July 1). Values: Working on problems that really matter. *Sustainable Lens* [Podcast]. Retrieved from http://sustainablelens.org/?p=1022

Green, D. (2016, April 17). Participatory documentary storytelling. *Sustainable Lens* [Podcast]. Retrieved from http://sustainablelens.org/?p=998

Heltne, P. G. (2008). Imposed ignorance and humble ignorance – two worldviews. In W. Vitek & W. Jackson (Eds.), *The virtues of ignorance: Complexity, sustainability, and the limits of knowledge* (pp. 135–149). Lexington: University Press of Kentucky.

Knowles, B. (2014, March 3). Changing mindsets. *Sustainable Lens* [Podcast]. Retrieved from http://sustainablelens.org/?p=625
Krumdieck, S. (2015, June 30). Transition engineering. *Sustainable Lens* [Podcast]. Retrieved from http://sustainablelens.org/?p=859
Laitner, J. A. (2014, March 28). Intelligent efficiency. *Sustainable Lens* [Podcast]. Retrieved from http://sustainablelens.org/?p=636
Latham, R., & McIvor, J. (2016, April 1). Value driven bikes. *Sustainable Lens* [Podcast]. Retrieved from http://sustainablelens.org/?p=992
Mann, S. (2011). *The green graduate: Educating every student as a sustainable practitioner*. New Zealand: New Zealand Council for Educational Research.
Mann, S. (2016). Computing education for sustainability: What gives me hope? *Interactions, 23*(6), 44–47.
Mann, S., Smith, L., & Muller, L. (2008). Computing education for sustainability. *ACM SIGCSE Bulletin, 40*(4), 183–193.
May-Sherman, G. (2015, June 26). Transition Oamaru. *Sustainable Lens* [Podcast]. Retrieved from http://sustainablelens.org/?p=869
Meyers, E. M., & Nathan, L. P. (2016, February). Impoverished visions of sustainability: Encouraging disruption in digital learning environments. In *Proceedings of the 19th ACM Conference on Computer-Supported Cooperative Work & Social Computing* (pp. 222–232). New York: ACM.
Nathan, L. P. (2012, July 14). Information systems for societal challenges. *Sustainable Lens* [Podcast]. Retrieved from http://sustainablelens.org/?p=325
Norman, D. (2014, February 6). Usability: Sustainability. *Sustainable Lens* [Podcast]. Retrieved from http://sustainablelens.org/?p=666
Norström, A. (2015, November 26). Nurturing social-ecological transformation. *Sustainable Lens* [Podcast]. Retrieved from http://sustainablelens.org/?p=942
Osborne, P. (2015, April 1). Valuing value. *Sustainable Lens* [Podcast]. Retrieved from http://sustainablelens.org/?p=827
Penzenstadler, B. (2016, January 7). Sustainable software engineering. *Sustainable Lens* [Podcast]. Retrieved from http://sustainablelens.org/?p=956
Pierce, J. (2012, May). Undesigning technology: Considering the negation of design by design. In *Proceedings of the SIGCHI Conference on Human Factors in Computing Systems* (pp. 957–966). New York: ACM.
Preist, C. (2013, May 24). Environmental impact of digital transformation. *Sustainable Lens* [Podcast]. Retrieved from http://sustainablelens.org/?p=473
Raghavan, B. (2016, February 21). Computing at the heart of culture change. *Sustainable Lens* [Podcast]. Retrieved from http://sustainablelens.org/?p=980
Remy, C. (2016, July 27). Loyalty to devices, not just brands. *Sustainable Lens* [Podcast]. Retrieved from http://sustainablelens.org/?p=1071
Robèrt, K. H. (2015, November 20). Strategic sustainable development. *Sustainable Lens* [Podcast]. Retrieved from http://sustainablelens.org/?p=938
Root-Bernstein, R. (2008). I don't know! In W. Vitek & W. Jackson (Eds.), *The virtues of ignorance: Complexity, sustainability, and the limits of knowledge* (pp. 233–250). Lexington: University Press of Kentucky.
Russell, D. (2016, July 8). Searching for sustainability. *Sustainable Lens* [Podcast]. Retrieved from http://sustainablelens.org/?p=1024
Schien, D. (2016, July 21). Footprints of digital infrastructure. *Sustainable Lens* [Podcast]. Retrieved from http://sustainablelens.org/?p=1030
Silberman, S. (2012, July, 12). joined-up thinking. *Sustainable Lens* [Podcast]. Retrieved from http://sustainablelens.org/?p=321

Silberman, M. S., Nathan, L., Knowles, B., Bendor, R., Clear, A., Håkansson, M., . . . Mankoff, J. (2014). Next steps for sustainable HCI. *Interactions*, *21*(5), 66–69. https://doi.org/10.1145/2651820

Stagl, S. (2007). Theoretical foundations of learning processes for sustainable development. *International Journal of Sustainable Development and World Ecology*, *14*(1), 52–62.

Thiel, P. (2015, November 30). values-based change agent. *Sustainable Lens* [Podcast]. Retrieved from http://sustainablelens.org/?p=929

Tomlinson, B. (2012, January 6). Dr Bill Tomlinson. *Sustainable Lens* [Podcast]. Retrieved from http://sustainablelens.org/?p=309

Toyama, K. (2016, January 1). Technology amplifies underlying human forces. *Sustainable Lens* [Podcast]. http://sustainablelens.org/?p=933

Vitek, B. (Ed.). (2008). *The virtues of ignorance: Complexity, sustainability, and the limits of knowledge*. University Press of Kentucky.

Vitek, B., & Jackson, W. (2008). Introduction: Taking ignorance seriously. In *The virtues of ignorance: Complexity, sustainability, and the limits of knowledge* (pp. 1–17).

Willard, B. (2005). *The next sustainability wave: Building boardroom buy-in (Conscientious Commerce)*. Gabriola Island, British Columbia, Canada: New Society Publishers.

Response 1a Sustainable HCI
From individual to system

Chris Preist

Roedl, Odom and Blevis (this book) focus on the role of interaction design in enabling more sustainable computing products – partly through the critique of existing unsustainable practices. This is 'sustainability in design' of digital technology – or the industry sector 'footprint' as Mann and Bates (this book) refer to it. Mann and Bates take the discussion beyond the footprint, to the 'handprint' of digital technology in a playfully provocative critique of sustainable HCI, drawing in diverse strands from the wider sustainability literature. Finally, Clear and Comber (this book) focus on the role of practice theory – a particular 'turn' in sustainable HCI research which has emerged as possibly the most significant response in the community to the critique that the focus on individual behavior and persuasive technologies was ill-founded.

All three papers emphasized the need for a transdisciplinary perspective and point to relevant discipline areas – and Mann and Bates 'lift' a framework from business studies and 'repurpose' it for use as a research ethics tool. I agree, but feel that we sometimes cast our net too narrowly and ignore the value of quantitative approaches from science and engineering (such as industrial ecology) in helping us answer key questions. Furthermore, drawing on social science, there has been a significant focus on observations of everyday practice in the HCI community. With a few exceptions (Hasselqvist et al., 2016), I find the insights these offer (that Clear and Comber explore) to be thought-provoking but partial. Like the 'individual behavior change' philosophy it has replaced, these still focus on the individual, albeit within the context of the social forces which constrain and shape them. If HCI is 'the relationship between computers and society' we will need to link more deeply with disciplines such as management science, economics, law and policy studies, which consider the ways in which social institutions shape practices of both individuals and collectives, including companies large and small. We also need a broader understanding of the systems in which we, and the digital services we use and design, sit. Understanding the barriers to possible strategies, and their consequences, both qualitative and quantitative, can help us understand more about what is possible.

As an example, let's consider practices around renewal and reuse discussed by Roedl, Odom and Blevis. Digital devices are regularly reused in the home by other family members, particularly children. However, as this happens, practices and expectations in the home change rapidly: children become used to, or even reliant on, such devices and are so 'inducted' into the adult world. Perhaps, in a couple of years they are no longer satisfied with an old model . . . Hence, reuse results in more "users". A similar phenomenon happens on a far larger scale globally: the market for reconditioned smartphones is estimated by Deloitte to be around $17 billion in 2017, and is largest in Africa and Asia. Devices have lives significantly beyond the "premium western" two years as they are reconditioned and moved around the world. However this is because, from a business perspective, these cheaper devices open opportunities for new markets among people who would otherwise not be able to afford them. More phone users – but less resource use than would have happened if they were sold new low-end phones. And that leaves 'us' with the classic western environmentalist dilemma: who are 'we' to deny them this opportunity that 'we' already have?

And here again, widening the system boundaries combined with quantification will help: Pargman and Raghavan (2014) have argued that we need to take a perspective of ecological limits in design, which are quantified in the science of planetary boundaries (Rockström et al., 2009). Is it possible to enable the whole of humanity to have access to digital services to the same level that we have in the western world (Preist & Shabajee, 2010)? Will a combination of 'circular economy' design principles, low energy technologies, and sustainable interaction design of digital services enable this within a certain small 'share' of energy and resources that the planet can sustainably provide? For a technology such as aviation, the answer is almost certainly no. But for digital services, which have a far smaller individual footprint and are currently shared over more of humanity, the answer just might be yes. And if the answer is no, then we can look at why; what innovation, and what restrictions on use, might be necessary to make it so? This can be thought of as a quantified version of backcasting of the kind that Mann and Bates advocate, a research program which is realistic and takes a globally egalitarian standpoint – but nonetheless which could appeal to more progressive parts of the IT industry.

References

Hasselqvist, H., Hesselgren, M., & Bogdan, C. (2016). Challenging the Car Norm: Opportunities for ICT to support sustainable transportation practices. In *Proceedings of the 2016 CHI Conference on Human Factors in Computing Systems* (pp. 1300–1311). New York: ACM.

Pargman, D., & Raghavan, B. (2014). Rethinking sustainability in computing: From buzzword to non-negotiable limits. In *Proceedings of Nordic Conference on Human-Computer Interaction (NordiCHI '14)* (pp. 638–647). New York: ACM.

Preist, C., & Shabajee, P. (2010, November). Energy use in the media cloud: Behaviour change, or technofix? In *Proceedings of IEEE International Conference on Cloud Computing Technology and Science (CloudCom 2010)* (pp. 581–586). IEEE.

Rockström, J., Steffen, W., Noone, K., Persson, Å., Chapin III, F. S., Lambin, E., . . . Nykvist, B. (2009). Planetary boundaries: Exploring the safe operating space for humanity. *Ecology and Society*, *14*(2), 32.

Response 1b Sustainability within HCI within society

Improvisations, interconnections and imaginations

Janine Morley

What is the place of HCI within society, and thereby, within sustainability? In their own ways, the chapters in this section take stock of this relationship and together they deliver a mixed status report: at once marked by a sense of progress and ongoing promise but also of frustration and limitation. In doing, they revisit the different roles that HCI has defined for itself with respect to sustainability. Roedl, Odom, and Blevis discuss the progress made in working with sustainability as a consideration *in* the design of digital products, namely, by aiming to prolong their longevity and challenge the speed of invention and disposal. In contrast, Clear and Comber focus on how to design *for* sustainable ways of living, more broadly defined. This familiar distinction (Mankoff et al., 2007; Hilty and Aebischer, 2014) of, on the one hand, "reducing computing's own footprint" and, on the other, supporting a "positive socio-economic transformation" is discussed by Mann and Bates, who lament what they see as a persistent lack of engagement within HCI to support such a societal transition.

When it comes to the sustainability implications that unfold as products are taken up and used, these approaches are not as distinct as they might appear. The separation of what technologies are *for* from their direct environmental footprint is somewhat artificial: that is, without having established and without sustaining some kind of role for such products within practices, there would not be a direct environmental footprint at all. However, I agree that sustainability in design and sustainability through design are very different in terms of their implications for HCI and the type of research it carries out. One approach remains within hitherto familiar co-ordinates, defining itself by reference to the design of digital products and the situated interactions users have with them. The other marks a potentially radical departure, taking as its focus not technologies themselves, but ways of living and processes of social change. In de-centering digital technologies and users, the challenge of sustainability is also emerging as a challenge, and opportunity, for HCI to define and re-define the distinctiveness of its work (Pierce et al., 2013). I am intrigued by the divergent futures this seems to imply for HCI, so in this short commentary I focus mostly on the chapters that discuss this challenge (Clear and Comber, Mann and Bates).

Improvisations

Clear and Comber review and elaborate a framework for design research to facilitate sustainable social practices, taking those practices as the unit of design. This draws on Kuijer et al. (2013) to advocate a process of necessarily radical change achieved through a "crisis of routine" and a series of "improvised performances". In doing, Clear and Comber point towards an agenda for "transdisciplinary design research" that moves away from an *a priori* computer-based framing of 'problems' and 'solutions'. If we extend their analysis and consider HCI research to consist in a set of practices, this response to the challenges of sustainability could also be seen to represent a crisis of routine for HCI, and the related development of new methods, frameworks and concepts, as a kind of improvised response to de-centering its long-established focus.

Mann and Bates also seek to re-define and extend the hitherto traditional boundaries of the field by arguing that HCI concerns the "relationship between computers and society". Whilst HCI is often thought of as an interdisciplinary field rather than a discipline in its own right, these moves imply a shifting balance between computer science and other disciplines like sociology and design research. This raises many questions. For instance, amongst other fields that also address the relationship between computers and society (in sociology, STS, organizational and Internet studies, and technology governance) what makes HCI distinctive? Is it an explicit aim to "design" (and re-design) aspects of this relationship? In other words, how do HCI researchers undertake to mediate the relationship between computers and society: Through what kind of research? And what does this imply for "improvisations" in the mixture of methods, theories, and objectives that are drawn from across connecting disciplines?

Interconnections

Building on Clear and Comber's discussion, the distinctiveness of HCI might be formulated and investigated as a question of how it, as a set of practices, is immersed in the massively varied and interlinked plenum of social practices that constitute society (Schatzki, 2012). For instance, Clear and Comber discuss the notion that "stakeholders" are "invested in the configuration of the various elements that make up everyday practices". That is, even though stakeholders are not always visible or present at sites of performance (for instance, when cooking is done, or when commuting takes place), they have nevertheless contributed to shaping those activities (perhaps, by providing certain kinds of products or investments in infrastructures) and have interests in sustaining (or altering) how practices are organized. Clear and Comber do not explicitly position HCI researchers as stakeholders in everyday practices, but others have, in asking how design and everyday practices interconnect as part of an ecology of practices (DiSalvo et al., 2013).

Indeed, it is worth asking this question more generally: in which practices might HCI *already* be considered a "stakeholder"? That is, how has HCI already made contributions to shaping the ways that practices beyond its own community are organized?

This is likely to implicate a broad array of practices in addition to those which are designed for governance and other research communities with which HCI interacts, including commercial innovation and production. Like HCI and the 'everyday' practices it often designs for, these practices are also part of society: not separate from or acting upon it, as if from outside. And through these 'indirect' interconnections with other practices HCI research may already be giving shape to the more and less sustainable futures that are emerging across these complex systems of practice.

This implies a reflexive awareness for those strands of HCI that are serious about contributing to sustainable transitions that engage with the unfolding implications of its work. This is not quite as straightforward as assessing all HCI research on sustainability criteria, as Mann and Bates suggest, since both low- and high-ranking research may have an equivalent (lack of) interconnection within this wider system of practice and thereby similar implications for sustainability. Instead, it implies a more thorough engagement with understanding, and potential innovation in, the ways that HCI research interconnects with wider practices in society, including other disciplines (through what it draws on and through its research 'outputs'). It implies greater clarity of what it might mean to do research and design at the interface(s) between 'society' and 'computers', whilst accepting that computers are part of society. It implies thinking about what ideas and principles about design and technology are being developed and how they might 'circulate' amongst other practices, and not simply as 'final' products based on experimental objects or lines of code. This may ultimately call for new kinds of imagination.

Imaginations

Clear and Comber suggest that, when it comes to the complexities, obduracies and multiplicity of stakeholders involved in social practices, design fictions may be a more promising output for HCI than technological 'solutions'. I am unfamiliar with this approach but I am intrigued by what kind of research it implies. By what methods are design fictions developed and then 'used'? To what extent do these fictions focus on technologies? In what ways can they be transdisciplinary? And are they pure fictional speculations or seen as "plausible future systems" that are "qualitatively possible, in terms of the possible co-emergence and interaction amongst the multiple elements that constitute them" (Tyfield et al., 2016, p. 3); that is, might they share qualities with scenarios developed to explore social futures? In short, what kinds of social and/or technological imaginations do they develop?

In sum, the chapters in this section together paint a portrait of a field that is, in part, re-imagining itself and improvising in its methods and theoretical foundations as it grapples with the challenges of sustainability. As it does so, there may be chances to interconnect with other disciplines in different ways, particularly ones that are also responding to the challenges of sustainability and re-imagining how their research may contribute to shaping social, and not merely technological, futures (Urry, 2016; Kuijer and Spurling, 2017).

References

DiSalvo, C., Redström, J., & Watson, M. (2013). Commentaries on the special issue on practice-oriented approaches to sustainable HCI. *ACM Transactions on Computer-Human Interaction, 20*, 1–15.

Hilty, L. M., & Aebischer, B. (2014). ICT for sustainability: An emerging research field. In L. M. Hilty & B. Aebischer (Eds.), *ICT innovations for sustainability*. Springer.

Kuijer, L., Jong, A., & Eijk, D. (2013). Practices as a unit of design: An exploration of theoretical guidelines in a study on bathing. *ACM Transactions on Computer-Human Interaction, 20*, 1–22.

Kuijer, L., & Spurling, N. (2017). Everyday futures: A new interdisciplinary area of research. *Interactions, 24*, 34–37.

Mankoff, J. C., Blevis, E., Borning, A., et al. (2007). Environmental sustainability and interaction. In *CHI '07 Extended Abstracts on Human Factors in Computing Systems* (pp. 2121–2124). San Jose: ACM.

Pierce, J., Strengers, Y., Sengers, P., et al. (2013). Introduction to the special issue on practice-oriented approaches to sustainable HCI. *ACM Transactions on Computer-Human Interaction, 20*, 1–8.

Schatzki, T. R. (2012). A primer on practices. In J. Higgs, R. Barnett, S. Billett et al. (Eds.), *Practice-based education: Perspectives and strategies* (pp. 13–26). Rotterdam, The Netherlands: Sense Publishers.

Tyfield, D., Zuev, D., Ping, L., et al. (2016). The politics and practices of low-carbon urban mobility in China: 4 Future scenarios. *CeMoRe Report*. Lancaster University.

Urry, J. (2016) *What is the future?* Cambridge, UK: Polity Press.

Photo Essay 3

Rooftop garden (2015). An urban rooftop garden in Shanghai illustrates the possibility of overly romanticizing an idyllic return to sustainable, simple life at scale – this garden occupies some of the most expensive real estate in the world and is not realistically available to all but the richest few. Reflection on Ekbia and Nardi: *Developing a political economy perspective for sustainable HCI*, and Penzenstadler: *Does sustainable HCI require different software engineering?* and Desjardins: *Reflections on longevity, unfinishedness and design-in-living.*

Eli Blevis

Part 2
Addressing limits

4 Every little bit makes little difference

The paradox within SHCI

Somya Joshi and Teresa Cerratto Pargman

Introduction

The Apollo mission[1] gave us the "blue marble" metaphor of Earth, which allowed for a semantic separation of the planet as a system and our own human agency within it. The subsequent call to "reconnect with the biosphere" (Folke et al., 2011) forms the grand narrative of the Anthropocene (Bonneuil & Fressoz, 2016), where we are encouraged and comforted by an alleged awakening brought on by new knowledge and reflexivity. This relationship between our planet and us (as something separate) is resonated within sustainable human-computer interaction (SHCI), where we find monitoring, measuring and incremental gains in knowledge and efficiency being put forward as solutions (Brynjarsdottir et al., 2012), rather than any call for systemic shifts in response to the finite resources and capacity of the planet to sustain life. In this chapter, we explore the inherent tensions and conflicts between the push for novelty and the drive to design digital tools and systems in the name of sustainability on the one hand, and the hard limits to consumption, development, and growth on the other hand. We do so by focusing on two key processes at play here. The first relates to the ways in which openness, collaboration, modularity, and longevity are translated into the technologies themselves (Blevis, 2007). The second relates to the social movements based on normative values that are catalyzed and nurtured via design processes. Here we draw on the body of work done on activism and design (Di Salvo, 2012; Fuad Luke, 2009) and argue that no attempt at achieving sustainability can be successful without the two processes being intertwined. In this chapter we argue that sustainability needs to be seen as a process rather than as a product or endpoint. This temporality, along with the interconnectedness of the endeavor, contribute to the conflicts, to the contradictions, to the tensions, and to what we call the paradox within HCI. We situate this paradox at the level of conflicting views of the world that get reflected in current design practices and intentions. It is one thing to design with the aim to challenge current paradigms of electronics consumption and waste that are operating within a specific market logic; it is another thing altogether to design with the aim of challenging such a market logic and thereby confronting "the economic system as a driver of unsustainable futures" (Nardi and Ekbia, this book).

The chapter is structured as follows: We provide a critical discussion on how sustainability as a concept has been framed in the wider literature (drawing on diverse disciplines outside of SHCI, ranging from political ecology to sustainable development, from economics to computation science). From here we narrow down our field of inquiry to address how sustainability has been appropriated within the HCI domain. Our intention here is to situate our own research work, to position it within this landscape, and to demonstrate how the shared metaphors (or lack thereof) translate into conflicts, paradoxes, and tensions within the design of technologies. We then discuss our methodology and empirical case in depth, to ground the discussion. Following from this we build upon current understandings of the role of sustainability principles within technology design. We conclude the chapter with a set of open questions regarding *what kind of world we want to sustain* and if we need to transcend the disciplinary boundaries of HCI as it stands today in order to do so.

The moving carousel of sustainability

"Sustainability is a process, not an endpoint."

(Silberman et al., 2014, p. 6)

"Sustainability is an active condition of problem solving, not a passive consequence of consuming less."

(Tainter, 2006, p. 93)

The way in which sustainability is framed within the literature ranges from a focus on resilience and vulnerability (Gallopin, 2006) to calls for social sustainability (e.g., Lehtonen, 2004), to approaches built around conceptualizations of justice (Dobson, 1996). Building on this, we see it as continuous transactional processes in which different forms of capital are mobilized, activated or formed (Joshi & Cerratto-Pargman, 2015; Cerratto Pargman et al., 2015). This particular view can be related to Tainter (2011), who emphasizes that "sustainability is not the achievement of stasis . . . it is rather a process of continuous adaptation, of perpetually addressing new or ongoing problems and securing resources to do so" (p. 33). In these continuous processes questions such as for instance "what will we give up?" (Nardi, 2013) are central to design practices caught between a push for novelty and the drive to design digital tools as well as the limits to consumption, development and growth.

The tension between what to give up also resonates in the terms "sustainability" and "sustainable development", which are often used interchangeably. The term "sustainable development", as referenced in the Brundtland report, is the result of a compromise between environment-first and economic development/social justice–first economic development/social justice-first proponents (Pargman & Raghavan, 2014). This struggle between prioritizing developing, fast-industrializing nation needs over the hard limits of planetary capacity to sustain life, results in a breakdown of the concept built around delusions, vagueness, and hypocrisy (related to promises of growth hitting against limits of ecology) as highlighted by Robinson (2004) as well as Pargman and Raghavan (2014).

Resilience thinkers such as Hajer et al. (2015) refer to the ineffective top-down initiatives (also referred to as "cockpitism") to promote sustainability, either via governmental directives or international agreements. They argue:

> We are currently operating outside both sets of boundaries, facing . . . human deprivation and environmental degradation: moving into the "safe and just space" will demand both far greater efficiency in resource use for meeting human needs, and far greater equity in its global distribution.
> (Hajer et al., 2015, p. 1654)

Thus they frame sustainability as an issue that can be addressed with greater efficiency and equity. If, as Toyama (2011) puts it, technology is just a tool that can be used to amplify human intent, what implication does this have for technology design aspiring towards sustainability in a limitless growth model? Answers to these questions emerge from the field of political ecology that took form in the 1960s and 1970s to offer a more critically inflected and social diagnosis of environmental change and the conflicts to which it often gives rise. In contrast to purely technical analyses of environmental change, political ecologists emphasized "political sources, conditions and ramifications" (Bryant, 1992, p. 13). To that end, they critiqued the way that science is often used to naturalize inequality and exploitative socio-natural arrangements. Debates concerning "political ecology" typically refer to the social and political conditions surrounding the causes, experiences, and management of environmental problems (e.g., Blaikie & Brookfield, 1987; Greenberg & Park, 1994; Zimmerer, 2000). In an early book on political ecology, Miller (1978) wrote:

> A primary contribution of a "political ecology" movement should be to demythologize this idealist mystification of the human/nature relationship [as adopted by economic exploitation] and to begin the construction of a new, holistic ethic . . . With nature and people increasingly viewed as having only commodity and exchange value, the acquiescence of science to that same perspective can only lead to a deepening dehumanization within society and to a further exploitation of nature.
> (Miller, 1978, p. 56)

Thus, we see a recurrence of the tension inherent within the framing of sustainability to balance human development with natural planetary capacity. What Miller (1978) was alerting us to was the mythology of the human-nature dichotomy. This relationship continues to be fraught with conflicting expectations and asymmetries of political power.

Moving away from this view, Pargman and Raghavan (2014), offer a very concrete definition of sustainability within this context, when they refer to it as, "*an absolute measure and an end-state in which the Ecological Footprint of humanity is below the regenerative bio-capacity of planet Earth*". This is a non-negotiable limit, regardless of political, economic or social interests. While we are sympathetic to this more radical interpretation of sustainability, we ask: Is sustainability an endpoint or a process that needs sustaining in and of itself? Can sustainability as a state ever be achieved without thinking of how it can be maintained and how we need to design our practices and tools to sustain that change over time?

Meadows et al. (2004) add to this understanding a range of scenarios for the global economy in their modeling study *Limits to Growth*. Their objective here was to propose how ecosystem limits placed limits on the global human economy. In their opinion the way sustainability has been framed (within industrial economic development terms) forces us between a resource and a pollution crisis. Nardi (2013) reinforced this argument by stating: "*humans remake themselves through technologies in intended and unintended ways – ways that often profit those in power*" (2013). This links back to the argumentation put forward by the political ecologists, questioning how value was constructed within such frames.

Thus we see from development discourse to economic models, from resilience studies to political ecology, sustainability as a concept has been fraught with tensions and conflicting worldviews. How are these tensions reflected in our attempts to address sustainability challenges through technology design? We turn our attention to this precise question in the segment below.

Sustainability within human-computer interaction

Blevis (2007) coined the term *sustainable interaction design* (SID) in an attempt to introduce a perspective on sustainability as a central focus of interaction design. A decade later, we realize that dealing with sustainability issues in the field of HCI is a more complex task than we thought it would be. The field is still characterized by significant differences in theoretical orientation, research methodology and practical objectives (Goldman, 2009). Such differences contribute to different and somewhat competing environmental discourses in HCI. Knowles et al. (2014) speak for instance of a reformist and a radical discourse that reflects profound differences in the way sustainability is approached both conceptually and in practice in the SHCI design space. While a reformist discourse on the environment relies on "ensuring continuance of the current standard of living to future generations, the radical discourse seeks answers to why we might want to sustain the world as is" (Knowles et al., 2014, p. 310) and it moves further away from the "conditions created by and supporting industrialism" (Knowles et al., 2014, p. 309). Assumptions underpinning these discourses are reflected in current understandings of what the design space of SHCI is and how we can operate on it.

This chapter revisits the original design principles, as put forward by Blevis, via the lens of our empirical case. In doing so, our intention is to provide critical insights into where the complexities, paradoxes and conflicts lie between design intentions and values on the one hand, and sociopolitical and ecologic limits within which we operate on the other hand.

Critiquing current approaches in SHCI as inherently self-limiting, Dourish (2010) underscores that "research in HCI for environmental sustainability has systematically ignored important areas for potential action" (p. 8) such as political and cultural contexts of environmental practice. For instance he observes that HCI has transformed the problems of sustainability into the cost-benefit trade-offs of rational actor economics, promoting sustainability as a "matter of personal morality rather than industrial regulation or political mobilization" (Dourish, 2010, p. 2). In this sense, Dourish speaks of "designing technologies of scale-making" (2010, p. 3) aimed at connecting people *through* their actions and their consequences, not on connecting people *to* their actions and their consequences, as it has been the case with the design of persuasive technologies. As such he calls for an approach of sustainability in HCI that attempts to move away from fostering environmental consumers to shaping environmental movements. "If we see the problem of environmental responsibility to be a problem of the ways in which people are linked together through their commitments, interests, and actions, this approach takes these connections as the primary focus of design attention" (Dourish, 2010, p. 7).

The last few years of SHCI have seen efforts to encourage design work that does not produce technological novelty, with discussions on, for example, of appropriation, maintenance, and repair (Huh et al., 2010), the "implication not to design" (Baumer & Silberman, 2011), "undesigning technology" (Pierce, 2012), and technology non-use (Baumer et al., 2014). Yet a tension persists among SHCI researchers that promote non-use, even when warranted by sustainability considerations, as it places them at odds with a central, if implicit, tenet of HCI at large. Knowles (2014) refers in this context to a material insecurity and survival anxiety coupled with a desire for change.

Resolving the tension between the "de-growth" implications of sustainability discourse and HCI's traditional promotion of invention, novelty, and innovation as ends in themselves will require the development of a nuanced, flexible, and sensitive discourse that might be called appropriate or responsible innovation (Edgerton, 2011; Grimpe et al., 2014). We see political economy as introduced in this book by Nardi and Ekbia (2018) as a key dimension of such discourse on appropriate or responsible innovation.

Methodology

This chapter draws on data that has been collected over the last three years (2014–2017) following the evolution of our case, that of Fairphone (FP). The data has been drawn at two levels, namely interactional and reflective data. Interactional data was collected via the website, blogs, and online documentation, as well as the critical voices emerging from the wider community of users and supporters on the FP Forum. The latter consists of early adopters, experts, users, designers, and partner organizations. Over 500 forum posts were scrutinized across parallel threads along with social media data from Twitter and Facebook that we followed since 2014. An interpretive approach (Walsham, 1995) was adopted in the analysis of this corpus. As we were interested in accessing the diverse interpretations of people in the field situations, we also conducted in-depth interviews.

The semi-structured interviews conducted iteratively with impact development and product strategy staff at FP constitute the reflective data collected. The interviews focused mainly on how FP regarded themselves as both a social enterprise and a competitive player in the market of consumer electronics. We were in particular interested in trade-offs that emerged when embarking on such a journey. The questions we asked focused on people's backgrounds, qualifications and affiliations to other movements, the historic reasons behind FP and its ideology. We also explored how decisions were frequently taken and what were the biggest challenges or risks as well their major roadblocks. Inquiries about how they perceived their real mandate in relation to the users and the real costs related to FP's ambition to maintain a high level of transparency, outreach, engagement and lifecycle planning. The personal use of the FP by one of the authors, as well as the commitment to sustainability issues in relation to electronics, informed the types of questions asked in the interviews. Each interview lasted between one hour and one hour and 30 minutes and was conducted through an online video-conferencing platform. Complementary email interviews were conducted to clarify and further develop the interviewees' statements.

The systematic analysis of the interactional and reflective data was conducted in the following manner: First, from the interview transcripts we identified a series of conceptual threads that we used as entry points in the reading of the blog, expert reports, forum debates and social media excerpts (the authors have followed FP's Twitter and Facebook flows regularly since they started the study in 2014). Second, these findings, which were obtained from a bottom-up analysis of both the interactional and reflective data, were in turn analyzed according to Blevis' (2007) sustainable interaction design principles. In the next section, we unpack the findings of our study.

The journey from hope to hype to hubris: the case of FP

What is FP?

FP started off as an awareness-raising campaign in 2010, mainly focused in the first instance on conflict minerals within the context of the smartphone industry. In the absence of a real alternative to point to, it emerged as a social enterprise in 2013, as the outcome of a crowdfunded campaign, designed to produce a truly 'fair' phone. By social enterprise, we refer to organizations and initiatives that harbor two goals: (1) to achieve social, cultural, community economic and/or environmental outcomes; and (2) to earn revenue.

We chose this case as it provides us a window into two worlds simultaneously: that of a social enterprise setting out to engineer and sustain a movement (based on changing relationships and behavioral practices within the domain of technology design and consumption); and that of a technical artifact designed to embody the lifecycle approach – built on the "fairware" principles (open hardware and software, modularity, conflict-free and fair in terms of workers rights, circular economy – from "cradle to cradle" approach). In other words this case provided us with the opportunity to dig deeper into our research question: *what role can the design of technologies play in achieving and maintaining sustainability within a world of limits?*

Objectives, assumptions and expectations

The FP social enterprise has been fundamentally driven by the following three objectives: (1) to develop a competitive product through a lifecycle design view, (2) changing consumption patterns and relationships with industrial partners (i.e., telecom and plastic providers), (3) opening up the black box of design for users and providing reflection on manufacturing practices that are often hidden by the impenetrability of electronic products. Below we address each of the design intentions of FP to better understand the motivations and drivers behind their choices in designing technologies for sustainable outcomes.

"The smartphone that is creating social change"

Just as DNA comes in pairs, we were informed in our discussion with the Head of Impact at FP, that the DNA of this social enterprise consists of two strands: on the one hand, it attempts to change the relationship between users and how they consume technology; on the other hand, it aims to use technology as an innovative tool to alleviate societal problems that emerge from its supply chain. We see here two key challenges standing in the way of realizing the goal of changing relationships between users and their technology fixes. That of scalability and the slow pace of change within the sustainability context. With regard to the former, a small outfit such as FP simply (to put it in their own words) *"can not afford to move the entire supply chain of production from China to Europe"*. This alludes to a set of compromises that shift their roadmap and milestones to impacts more graspable, small-scale and localized in the early stages of development. With regard to the latter challenge, in an industry and market such as that of smartphones and more broadly consumer electronics, the rate of change in product development feeds into an expectation of heightened novelty seeking. Within this environment, we ask, what does it entail to engage in a slow, deliberate march towards the attainment of sustainability goals, most of which are not immediately apparent (i.e., changing worker relationships, acceptance of standards and regulation with regard to conflict-free mining, recycle-reuse impacts)?

"Long-lasting, easy repair modular phone"

Be it manifest in the urban mining workshops organized routinely or the e-waste reduction efforts in Ghana, the attempt here is one of creating conditions for debates around sustainability. Urban mining emerges from this context as a way to change the existing imbalance, by the extraction of minerals from existing products. The aim is to dismantle gadgets that have reached their end of life to uncover what's within. The idea is that once users have unscrewed the back cover of their technological artifact (in this case a smartphone), and identified the components, the urban mining workshop would aim to unravel some of the phone's hidden stories. From pollution and extremely dangerous working conditions to child labor, one can learn that a number of mining-related practices desperately require improvement. The approach adopted by mainstream players within this context is to hide this inconvenient body of knowledge behind, sealed, glued and proprietary locked-in devices, where the design serves as an impenetrable casing which keeps all unpalatable, guilt-inspiring footprints of our consumption behaviors neatly out of sight (and hence out of mind).

"Putting working conditions and the environment first"

Framing itself as a social enterprise, FP opted from the start for a mandate that is more rooted in the discourse of sustainability than in any desire to be industry leaders in the smartphone domain. This has translated into a set of compromises or trade-offs. Transitioning from a young start-up to a credible market player today, it has found itself in the position where it is unable to diversify its product range, or maximize much needed profits in order to grow, but instead keeps a steady focus on the goal at hand, which is to promote fairness and quality control of the existing artifact. In the words of the CEO of the organization, *"Our mission is to create momentum to design this future. We started by making a phone to uncover production systems, solve problems and use transparency to invite debate about what's truly fair. We believe that these actions will motivate the entire industry to act more responsibly."* Two years down the line: *"This is where we still are. Our business model hasn't changed, nor has our product focus – we will still concentrate on phones, and won't branch out into other consumer electronics like laptops or tablets."* However such an expectation makes us wonder about the measurable impacts emerging from this choice to prioritize fairness over profits or growth. In particular we wonder how this translates into tangible changes within the established industry of consumer electronics, where it remains to be seen if the model of FP serves as an inspiration for change or a niche alternative which leaves no dent in their operations.

Revisiting sustainable interaction design principles

In this segment of the chapter we address each of the sustainable interaction design principles, as put forward by Blevis (2007), in light of our empirical findings.

Invention & disposal

Why design a phone in the first place? This is a question we run into when discussing sustainability and the consumption of technology in the same breath. In the case of FP the explicit articulation of their position is: *"Designing a phone is about more than developing a consumer product. The FP is a means to spark discussions and debate around the fairer production of mobile phones*." But what about the disposal and waste engendered by update cycles of software and hardware? By moving from FP1 to FP2, isn't there an inherent link between the new promises offered and the older promises betrayed? In Europe alone, it has been estimated that over 1.6 billion old phones are still in circulation, even though they are no longer actively used by their owners. Out of all these phones, only about 7% are properly recycled. In response to this FP launched an initiative to safely reuse and recycle the old phones produced by them. They also provided an unofficial space on their forum encouraging their community members to sell and reuse old models.

Promoting renewal & reuse

The above principle, in relation to longevity, resonates within the FP case study and in some ways forms the very raison d'etre for the movement. While the social enterprise has built itself around the principles of longevity and reuse, we found similar frustrations as the one voiced here regarding the lack of build quality:

> The FP2 is really disappointing. And this is not because I expect the newest state of the art technology . . . but I expect a phone that I can work with. I really paid a lot more . . . but I did that, because I believe that things should be as expensive . . . in order to be socially and ecologically sustainable. But for that money I wanted to get a mobile phone that should be good for the next years. Now after six months, my phone freezes and reboots whenever it feels like it. This is a pity because I hoped that I can tell everybody how great it is to have this phone because it is sustainable and it works.

The consumer electronic industry in general (but specially the smart mobile phone sector) is characterized by rapid software and hardware update cycles and forced obsolescence. Thus the choice to adopt a lifecycle perspective posed a critical compromise for the FP community, as it struggled to keep pace with the external market and environment within which it operated.

Promoting quality & equality

One early adopter of FP expressed his concern about the gap between rhetoric and reality, with regard to the ethical use of labor (equality) in the production of the phone:

> You have already published the sources of the materials and explained why you have chosen each source. So I know that my mobile phone doesn't support conflicts. But you haven't told us much about the situation of the workers who produce the FP. I know that it's really difficult to check all your partners but I'm not able to trust "pure words". You talk about transparency being a

crucial part of your "journey". Everybody can describe goals but this is what other companies (can) do. Proof and data are really important, too, not just "newspaper announcements". I want to support you but on the other hand I don't want to have any doubts left when I use my phone.

We see from the above critique that the community that supports FP (ranging from early supporters who crowdfunded the initiative to recent adopters) uses the transparency and openness of the social enterprise to make strong demands. These in turn have to be responded to if FP is to engage its end users in any meaningful way. Be it the decision to follow the entrenched supply chain to China where labor conditions are 'unfair' by any standards acceptable in the Global North, or the inability to trace minerals beyond a point (both geographically an in terms of degree of certitude), FP has come under a lot of criticism by its own community to live up to the expectations it set. For instance FP publishes transparently on its website the occasions where working hours for laborers in China exceeded 60 hours/week. The same was true of days off. While the employees generally have one day off in seven, there were also occasions where they did not get a day of rest. By raising awareness on such issues, the intention is to create a momentum for change. However does this contribute to sustainability? Is there a link between the ready availability of such labor (at lower costs and fewer benefits for the workers) and the scale of production and consumption today? Are we able to sacrifice longevity and upgrade devices at such an alarming rate of frequency precisely because we are enabled to do so in an unequal society that exploits certain segments to feed the hunger and desires of others?

While we applaud FPs attempts at opening the black box of design and asking such unpalatable questions around fairness and equity, we feel it is inadequate in achieving sustainability in the radical sense. Again, as an implication for design, what is needed is less "effecting change from within" and more systemic paradigm shifts that disrupt the very global supply chains we have come to take for granted, that are built on political fault lines of power and exploitation.

De-coupling ownership & identity

One of the principles for fair design at FP is that it should "... be a living object, evolve together with its owner". This resonates with what Nelson and Stolterman (2003) call ensoulment: "Ensouled things imply well-cared for things, looked-after things, durable and enduring things" (Blevis & Stolterman, 2007). Especially the material qualities of the Fairphone (e.g., non-conflict minerals, recycled plastic, technical specifications) and augmentation through inscriptions on the back and inside (battery and chip set) which state "yours to open, yours to keep" further cement the link with identity. This we argue enables the community of supporters to persevere through their identification and celebrate their ownership of an FP on ethical/sustainability grounds. This furthers longevity of the artifact. When coupled with reuse and repair, the device's life can be further enhanced. At the same time this also serves in a critical sense to reverse the direction of de-coupling – when FP supporters identify too closely with the brand of sustainability rather than the underlying functionalities themselves, leading to the support for upgrades and new iterations of the model.

Conclusion

> One of the problems with the 'greener gadgets' version of sustainable design is that the result is mostly still more stuff – hopefully less harmful stuff, but rarely just less stuff. Is it possible to use the practice of designing to eliminate stuff?
> (Tonkinwise, 2012, p. 115)

However, where they have been struggling is in their ability to scale the disruption to current supply chains and consumption trends on a systemic scale. Their need to 'play the game' in order to provide credible, current alternatives (a mandatory requirement for them to be able to offer support and services). This limits their imagination and urgency to change existing unsustainable practices. As an illustration of this is the recent FP attempt at modularity with a new camera upgrade. Their efforts are admirable in that they demonstrate modularity, long life and continued use of the same device, rather than rapid update cycles resulting in forced obsolescence. However the incremental nature of updates and modules changing is limited in imagination to more systemic shifts that question the very need for keeping pace with an unquenchable thirst for new and more functions. Opting for sufficiency over efficiency here, would seriously jeapodirse the market attractiveness of their product.

As we survey the landscape today and find the ruins and shells of similar ideas (such as Google's Ara project), FP emerges as the only case that has gone from being a start-up and activism-based organization to a self-sustained, credible market player, challenging and disrupting the discourse on sustainability and consumption. It would not exist today were it not for the chord it has struck with its consumer base, nor without its market robustness. However how does an initiative such as FP ensure that it is no longer needed, by changing procurement, production, consumption and disposal paradigms at a wide-enough scale to exit (much like the metaphor of development aid removing itself in the face of renewed capacity and sustainability)?

In this chapter we have looked under the hood of a technology design initiative to better understand the conflicts, tensions and paradoxes that arise when seeking sustainability as an outcome.

First, we have demonstrated how there is a lack of shared understanding around the non-negotiable limits of the ecosystem we inhabit. This results in an incremental approach, where 'every little bit counts' is accepted parlance. We have argued against this by calling for more radical systemic shifts within our social, political and economic spheres, as underscored by the de-growth movement and the research on political economy conducted by Nardi and Ekbia (2018).

Second, we have demonstrated via our case how frustrations and tensions arise when no thought is given to the post-design/post-event phase of sustaining engagement and adaptation to changing environments. The temporality of sustainability and need to sustain change is often ignored when designing interventions within HCI.

Third, we have argued that there is a false dichotomy inherent in the nature/humanity rhetoric, where human societal needs are seen in isolation of natural limits. Within SHCI, we have asked, what role can the design of technologies play in achieving and maintaining sustainability within a worlds of limits? By *making more stuff* as it were, are we really creating sustainable outcomes in a future of scarcity? Rather we recommend focusing our attention and resources more on shifting entrenched patters of identity, consumption, and political power that we have come to take as given. While a tall order no doubt, we argue this is a more honest approach than designing apps to monitor our footprints or add modularity as an afterthought design feature within a device feeding the hunger of consumer electronics. We fully support the call for more radical activism within the SHCI community that moves away from designing for raising awareness or efficiency in consumption, towards designing for worlds where different relationships are possible, based on longevity, reparability, equity, fairness or where different relationships need to be envisioned based on "the construction of a new, holistic ethic..." (Miller, 1978, p. 56) and a new political and economic order as suggested by Nardi and Ekbia (2018).

By presenting the often conflicting and divergent framing of sustainability, both within HCI and outside of it, we have shown how conflicting worldviews cohabit within a field that is concerned by the sustainability of our planet and driven by design practices that overlook the political economy dimension of design.

The diversity of voices around this thematic also alerts us within the SHCI community of the need to learn from other disciplines working with sustainability – to broaden our repertoire of metaphors and choices, of expertise and perspectives. Finally, we have shown via our empirical case how the best intentions towards the attainment of sustainability are challenged and at times thwarted when sustainability itself is treated as an endpoint or product. We have argued for the need in our field to connect with other disciplines such as political ecology (Miller, 1978), social ecology (Stokols et al., 2013) and political economy (Nardi and Ekbia). These disciples in particular remind us that we are part and parcel of the problem of how we think about sustainability issues and how we can act accordingly. Thereby they confront us to our own blind spots in our design and research practices (Cerratto Pargman and Joshi, 2015) and stimulate us to enact a paradigm shift that HCI needs in order to be able to rethink established and legitimized practices whilst envisioning new sustainable HCI principles, frameworks and methodologies.

Note

1 www.nasa.gov/content/blue-marble-image-of-the-earth-from-apollo-17

References

Baumer, E., Ames, M., Brubaker, J., Burrel, J., & Dourish, P. (2014). Refusing, limiting, departing: Why we should study technology non-use. In *Proceedings SIGCHI Conference on Human Factors in Computing Systems, (EA'14)* (pp. 65–68). New York: ACM.

Baumer, E., & Silberman, S. (2011). When the implication is not to design (technology). In *Proceedings SIGCHI Conference on Human Factors in Computing Systems (EA'11)* (pp. 2271–2274). New York: ACM.

Blaikie, P., & Brookfield, H. (1987). *Defining and debating the problem: Land degradation and society*. London: Routledge.

Blevis, E. (2007). Sustainable interaction design: Invention & disposal, renewal & reuse. In *Proceedings of the SIGCHI Conference on Human Factors in Computing Systems (CHI '07)* (pp. 503–512). New York: ACM.

Blevis, E., & Stolterman, E. (2007). Ensoulement and sustainable interaction design. In *Proceedings for International Association of Society of Design Research Conference* (pp. 1–23).

Blevis, E., Youn-kyung, L., Roedl, D., & Stolterman, E. (2007). Using design critique as research to link sustainability and interactive technologies. In *International Conference on Online Communities and Social Computing* (pp. 22–31). Berlin, Heidelberg: Springer.

Bonneuil, C., & Fressoz, J. (2016). *The shock of the anthropocene: The earth, history and us*. Scotland: Verso Books.

Brundtland Report. (1987). *Report of the world commission on environment and development: Our common future*. Retrieved from www.un-documents.net/our-common-future.pdf

Bryant, R. (1992). Political ecology: An emerging research agenda in Third-World studies. *Political Geography, 11*(1), 12–36.

Brynjarsdottir, H., Håkansson, M., Pierce, J., Baumer, E., DiSalvo, C., & Sengers, P. (2012). Sustainably unpersuaded: How persuasion narrows our vision of sustainability. In *Proceedings SIGCHI Conference on Human Factors in Computing Systems (CHI '12)* (pp. 947–956). New York: ACM.

Cerratto Pargman, T., & Joshi, S. (2015). Understanding limits from a social-ecology perspective. *First Monday, 20*(8). Retrieved from http://journals.uic.edu/ojs/index.php/fm/article/view/6125/4844

Cerratto Pargman, T., Pargman, D., & Nardi, B. (2016). The Internet at the eco-village: Performing sustainability in the 21st century. *First Monday, 21*(5). Retrieved from http://pear.accc.uic.edu/ojs/index.php/fm/article/view/6637/5515

DiSalvo, C. (2012). *Adversarial design*. Cambridge, MA: MIT Press.

Dobson, A. (1996). Environment sustainabilities: An analysis and a typology. *Environmental Politics, 5*(3), 401–428.

Dourish, P. (2010). HCI and environmental sustainability: The politics of design and the design of politics. In *Proceedings of the 8th ACM Conference on Designing Interactive Systems (DIS '10)* (pp. 1–10). New York: ACM.

Edgerton, D. (2011). *Shock of the old: Technology and global history since 1900*. Profile Books.

Folke, C., Jansson, Å., Rockström, J., Olsson, P., Carpenter, S. R., Chapin, F. S., . . . Westley, F. (2011). Reconnecting to the biosphere. *Ambio, 40*(7), 719–738. http://doi.org/10.1007/s13280-011-0184-y

Fry, T. (2009). *Design futuring: Sustainability, ethics and new practices*. ☐Bloomsbury Academic.

Fuad-Luke, A. (2009). *Design activism: Beautiful strangeness for a suitable world*. London: Routledge.

Gallopín, G. (2006). Linkages between vulnerability, resilience, and adaptive capacity. *Global Environmental Change, 16*(3), 293–303.

Goldman, E. (2009). Three environmental discourses in human-computer interaction. In *Proceedings of the SIGCHI (EA'09)* (pp. 2535–2544). New York: ACM.

Greenberg, J., & Park, T. (1994). Political ecology. *Journal of Political Ecology, 1*(1), 1–12.

Grimpe, B., Hartswood, M., & Jirotka, M. (2014). Towards a closer dialogue between policy and practice: Responsible design in HCI. In *Proceedings of the SIGCHI Conference on Human Factors in Computing Systems* (pp. 2965–2974). New York: ACM.

Hajer, M., Nilsson, M., Raworth, K., Bakker, P., Berkhout, F., De Boer, Y., Rockström, J., Ludwig, K., & Kok, M. (2015). Beyond cockpit-ism: Four insights to enhance the transformative potential of the sustainable development goals. *Sustainability, 7*(2), 1651–1660.

Huh, J., Nam, K., & Sharma, N. (2010). Finding the lost treasure: Understanding reuse of used computing devices. In *Proceedings of the SIGCHI Conference on Human Factors in Computing Systems* (pp. 1875–1878). New York: ACM.

Joshi, S., & Cerratto Pargman, T. (2015). On fairness & sustainability: Motivating change in the networked society. *Proceedings of ICT4S*, Atlantis Press, Copenhagen.

Knowles, B. (2014). Rethinking Plan A for sustainable HCI. In *Proceedings of the SIGCHI Conference on Human Factors in Computing Systems* (pp. 3593–3596). New York: ACM.

Knowles, B., Blair, L., Walker, S., Coulton, P., Thomas, L., & Mullagh, L. (2014). Patterns of persuasion for sustainability. In *Proceedings of DIS* (pp. 1035–1044). New York: ACM.

Lehtonen, M. (2004). The environmental – social interface of sustainable development: capabilities, social capital, institutions. *Ecological Economics, 49*(2), 199–214.

Meadows, D., Randers, J., & Meadows, D. (2004). *Limits to growth: The 30-year update*. Chelsea Green Publishing.

Miller, A. S. (1978). *A planet to choose: Value studies in political ecology*. Pilgrim Press.

Nardi, B. (2013). The role of human computation in sustainability, or, social progress is made of fossil fuels. In *Handbook of human computation* (pp. 1011–1020). Springer.

Nardi, B., & Ekbia. (2018). This text.

Nelson, H., & Stolterman, E. (2003). *The design way: Intentional change in an unpredictable world: Foundations and fundamentals of design competence*. Educational Technology.

Pargman, D., & Raghavan, B. (2014). Rethinking sustainability in computing: From buzzword to non-negotiable limits. In *Proceedings of NordiCHI'14* (pp. 638–647). New York: ACM.

Pierce, J. (2012). Undesigning technology. In *Proceedings of the SIGCHI Conference (CHI'12)* (pp. 957–966). New York: ACM.

Robinson, J. (2014). Squaring the circle? Some thoughts on the idea of sustainable development. *Ecological Economics, 48*(4), 369–384.

Silberman, S., Nathan, L., Knowles, B., Bendor, R., Clear, A., Håkansson, M., . . . Mankoff, J. (2014). Next steps for sustainable HCI. *Interactions, 21*(5), 66–69.

Stokols, D., Lejano, R., & Hipp, J. (2013). Enhancing the resilience of human – environment systems: A social ecological perspective. *Ecology and Society, 18*(1), Article 7.

Tainter, J. (2006). Social complexity and sustainability. *Ecological Complexity, 3*(2), 91–103.

Tainter, J. (2011). Resources and cultural complexity: Implications for sustainability. *Critical Reviews in Plant Sciences*, *30*(1–2), 24–34.

Tonkinwise, C. (2012). Sharing trust: Tasteful designs of social systems. *Trust Design: Public Trust*, *4*, 8–14.

Toyama, K. (2011). Technology as amplifier in international development. In *Proceedings of the iConference* (pp. 75–82). New York: ACM.

Walsham, G. (1995). Interpretive case studies in IS research: Nature and method. *European Journal of Information Systems*, *4*(2), 74–81.

Zimmerer, K. (2000). The reworking of conservation geographies: Nonequilibrium landscapes and nature-society hybrids. *Annals of the Association of American Geographers*, *90*, 356–369.

5 Developing a political economy perspective for sustainable HCI

Bonnie Nardi and Hamid Ekbia

Introduction

Since 2007, when sustainability was introduced as a key theme in the CHI community (Blevis, 2007), a remarkable body of work has emerged. In a newly critical spirit, HCI researchers have undertaken analyses of topics such as HCI's consumerist bent, its lack of attention to planned obsolescence and e-waste, and its reliance on narrowly conceived user-centered design and undertheorized notions of "the user" (see e.g., Blevis, 2007; DiSalvo et al., 2010; Maestri & Wakkary, 2011; Mota, 2011; Knowles et al., 2013; Dillahunt et al., 2014; Pargman & Raghavan, 2014; Hazas, 2015; Toyama, 2015). Research on reuse, repair, repurposing, redistribution, resilience, long-cycle disposal, multi-lifespan design, conservation of resources, collapse informatics, computing within LIMITS, simple living, and undesign has steadily accumulated, offering compelling new directions for the field (see e.g., Huang and Truong, 2008; Strauss and Fuad-Luke, 2008; Woodruff et al., 2008; Wong, 2009; Friedman & Nathan, 2010; Huh et al., 2010; Kuznetsov & Paulos, 2010; Baumer & Silberman, 2011; Pierce, 2012; Håkansson & Sengers, 2013; Lomas et al., 2013; Tomlinson et al., 2013; Bellotti et al., 2014; Dillahunt, 2014; Jackson, 2014; Gui & Nardi, 2015; Patterson, 2015; Remy & Huang, 2015; Chen, 2016; Franquesa et al., 2016; Qadir et al., 2016; Hazas et al., 2016; Sabie et al., 2016). These achievements are impressive for a decade's worth of effort.

Yet a paradox underlies this stream of research. An "elephant in the HCI room" stands before us, and that elephant is the political economy (Ekbia & Nardi, 2015). We know the economy is all-important, yet it does not figure in any principled way in our theories, practices, and designs. The next step for SHCI is, in our view, to situate research within a more sophisticated analytical perspective that theorizes and incorporates political economy. Current environmental problems are unambiguously the outcome of human economic activity. No one disputes that, yet we tend to shy away from confronting the economic system as a driver of unsustainable futures.

To have impact, we must inject our research with sociopolitical analysis. To say more about obsolescence, for example, it's important to pose questions about the properties of an economy that pushes us to consume. Designed obsolescence is not new; critics were complaining about it in the 1950s (Packard, 1960). Why is it still a problem? How does obsolescence affect our designs? Proposals for repair, reuse, and undesign expose the contradiction of an economy feeding off incessant growth and prodding us to throw away purchased items and replace them with new ones as soon as we can. Thoughtful plans for conservation, slowing, sharing, restraint, thrift, DIY, and simplicity must consider that the global economy – the most powerful force in our lives – actively squelches practices that do not promote consumption. "Economic growth" is the mantra of every government, not just corporations. In the bitterly contested Brexit vote, for example, both sides were in complete agreement that economic growth was the desired outcome.

Increased growth flies in the face of sustainability, which requires that we consume less not more, given the physical limits of Earth in resources and capacity to absorb and process waste (Pargman & Raghavan, 2014). With the understanding that people in poorer nations and social classes need access to the necessities of life, it is still true that aspirations for western lifestyles, and, more importantly, our own habits of consumption in the "overdeveloped" word, are not sustainable from a thermodynamic point of view (Daly, 1991; Booth, 1998; Hornborg, 2014; see also D'Allesandro et al., 2008, Jackson, 2009). Can we in SCHI do anything about such a massive problem?

It seems that we *must* do something if we are not to give up on future generations. SCHI is not merely a fascinating set of research topics; it confronts the signature contradiction of our era, whose impacts none of us can escape. Kaptelinin (2016) argues that HCI should develop an existential perspective that addresses primary, foundational human predicaments. We believe that sustainability is one such topic. There is everything at stake – not just academic careers, but our lives, and the lives of those who will come after us.

As members of the HCI community, we are in some ways uniquely positioned to engage the contradiction of political economy by virtue of a combination of several qualities we do not see in other research communities. These include the can-do ethos of engineering practice, the richness of perspectives from true interdisciplinarity, and commitment and openness to new ideas. Our objective in this chapter is to suggest that we build on these strengths and do what HCI has always done – i.e., "reach out", as Jonathan Grudin put it in his seminal 1990 CHI paper, to the next level of analysis. Areas such as ICTD, crisis informatics, and collapse computing have broadened the HCI purview. The next level of analysis is, in our view, the political economy.

Engaging this level is no easy task in a field in which the corporate presence is so strong. Many members of the HCI community work in industry. Industry is wedded to growth and has no agenda of long-term sustainability in the sense of defining commerce as an activity responsive to Earth's limits. Much of the work we do in HCI benefits the companies, and is indeed paid for by them – this is the realpolitik of the economy to which we belong, and our own contradiction and existential quandary. But we do not think we are tilting at windmills in continuing to develop SHCI and continuing to expose and address the paradox of, on the one hand, a political economy premised on growth, and, on the other hand, a commitment to sustainability. Many individuals who work in industry are well aware, as citizens and parents, if not as corporate employees, of the environmental precipice on which we are collectively teetering. They are aware of the science, aware that the future of their children and grandchildren is the endgame we play with Earth's limits. We can and should task ourselves with finding ways to develop and sustain an HCI conversation that investigates and commits to societal concerns beyond profit. This very volume is evidence of our willingness to do that. Developments in crisis informatics, collapse computing, ICTD, computing within LIMITS, and of course SHCI itself, demonstrate this commitment. These areas of HCI are relatively new, and we are encouraged by this "societal turn" that speaks to humanity's deepest problems within the context of the wider sociopolitical system.

What to do?

Yet we often founder on the specifics of the question, "OK, so what do I do?" We suggest three broad, general strategies for going about SHCI research with a political economy inflection. The strategies overlap, and we break them apart simply to isolate key emphases. The strategies actually presuppose one another; any project will start with the first, and may involve the other two.

1. The first strategy is to *select important research problems and treat them as aspects of the political economy*. We cannot emphasize this primary orienting move enough. Just asking a good question can be a powerful act. The answers to the important questions will be difficult, provisional, ambiguous, and maybe not too popular in some cases. The questions themselves may be controversial. But if we stop before we start, we cannot move forward. It is encouraging that every author and editor of this volume has chosen to pursue questions that challenge the mainstream. For example, Roedl et al. (this volume) observe that even critical discourse in HCI that examines obsolescence is "overwhelmingly focused on what design creates, rather than what it displaces and destroys. [There is] . . . a lack of rigorous engagement with the political economic dimensions of obsolescence." They observe that we must ask fundamental questions such as whether capitalist economies can even survive without planned obsolescence. Roedl et al. mention the importance of selecting good research questions: "Instead of creating new commodities and tools for consumption, interaction designers might instead devote their efforts to designing technologies that address basic human needs, as in the fields of HCI4D and Collapse Informatics" (Tomlinson et al., 2013). They point to the societal turn in HCI as promising and productive.

Why don't more people choose a path of seriousness, of enacting a research agenda with the potential for societal-level impact? We understand that getting tenure may involve the pragmatics of avoiding the complexities of political economy, yet we still see that many studies in crisis informatics, collapse informatics, computing within LIMITS, and ICTD involve young investigators – a heartening observation! After tenure there is no excuse. An important role for this volume, is, in our opinion, to encourage HCI researchers to reflect on the long-term impact of their work, and engage questions that matter for humanity, given the precarity of our current circumstances. Knowles and Ericksson (2015) write of the difficulties of doing so – once an understanding of the scale of the problems is accepted, we may feel "deviant and guilt-ridden" as we attempt to grapple with problems others studiously turn away from. Not only are we deviants, we must confront our own complicity in creating the problems. This faceoff can be uncomfortable and confusing. A common response to feelings of hopelessness is to fashion research that focuses on helping one small group of individuals directly, rather than tackling larger systemic questions that must necessarily engage the political economy, the dynamics of growth, and monstrous problems such as climate change.[1]

Clear and Comber (this volume) write that often in HCI the choice is to "promot[e] incremental change over the systemic change that is required". This is not a new theme in HCI or even one confined to SCHI; Kirman et al.'s satirical vision of a dystopian HCI future depicted a field that

> [devoted itself to] huge amounts of seemingly incremental user experience research which focused on generating minor improvements to interfaces. . . [that] had the specific result of encouraging humans to spend less time thinking about the fact that they were using technology, and more time. . . [c]onsuming.
>
> (2013)

It is not a new theme, but it is one that bears continual repetition and elaboration (as in Clear and Comber's cogent essay), because, to some extent, incrementalism is baked into normal HCI practice. We find this incrementalism especially visible in the rhetoric of "users". Clear and Comber (this volume) observe that framing issues at the level of individual users leads us to "ignore questions about the roles and responsibilities of . . . important social actors such as governments, institutions, and corporations, [including] their policies, regulations, and ideas of corporate responsibility in which our unsustainable ways of life are embedded. . . ". Governments, institutions, and corporations can only be brought into analysis through principled consideration of political economy.

Choosing problems that matter at the level of the political economy is likely to call into question not only "users", but a raft of HCI orthodoxies such as user-centered design, participatory design, and practice-based approaches. Questions arise not because these approaches are wrong, but because they nudge us too soon toward ordinary users to whom we can gain easy access. For example, in California, 80% of the state's water is allotted to industrial agriculture. Within the remaining 20%, only a small fraction goes to individuals; the rest is for government and institutional use. A typical "user-centered" HCI approach might be to build an app to monitor individual water usage. It would be easy to do a user study of this app! However, enabling people to save water in their homes and gardens will do essentially nothing to conserve water in California. Attempting to influence users has no traction for this problem. A better approach would be to consider how to change agricultural practice.

If we *start* with political economy, we might turn toward "rock the boat" approaches, as Ericksson and Pargman (this volume) call them. One example is computational agroecology which begins with a political economy analysis of the unsustainability of industrial agriculture, and then proceeds to consider computing solutions (Raghavan et al., 2016). User-centered approaches begin with users, bypassing deeper analysis of the sociopolitical systems in which user practice is enmeshed. Remy and Huang (this volume) observe, "[W]hen developing concepts that seek to raise awareness, change behavior, or promote knowledge, typically SHCI has often focused on the most visible stakeholder – the user – and neglected other parties." If we move too quickly to current practice, to users and their "needs", to user participation, we diminish the possibility of more radical approaches that question the foundations and structures of the political economy and our participation in that economy.

Roedl et al. (this volume) note, "[T]he underlying economic ideology within which design is intertwined . . . may be difficult to see." It is, without doubt, a difficult move to account for political economy in our work. Remy and Huang (this volume) ask: "What if one of the reasons for the difficulty . . . is that the other stakeholders . . . have not even been identified yet?" Ekbia and Nardi cast the problem somewhat differently, arguing that we know the economy and its powerful stakeholders are pertinent, but we treat them as the elephant in the room (2015). If we continue to do so, breakthroughs will, in all likelihood, elude us. Clear and Comber (this volume) remind us that important research topics are being bypassed: "[Less] research [in HCI] has focused on . . . transport, water conservation, thermal comfort, food (agriculture, supply and demand), and electronic waste." These topics invite treatment at the level of the political economy.

Selecting research problems should include studying populations that act on the fringes of society, or live on the fringes of society, such as the underclass in the US, social movements such as Transition Town or Steampunk, residents of ecovillages, and even survivalists who take extreme positions on the possibility of future scarcity. We shield ourselves from critiques of the everyday narratives of economic growth, progress, and infinite possibility (see Erickson and Pargman this volume) by turning too soon to ordinary, established practice. Those on the margins have been forced (the underclass) or have chosen (ecovillage residents, survivalists) to consider the political economy and its implications. These people merit more of our attention. Perhaps it is not surprising that SCHI has considered such groups more than other subfields of HCI (e.g., Nathan, 2008; Håkansson & Sengers, 2013; Cerratto Pargman et al., 2016). We hope to see more such work.

Computational agroecology: an example

The idea that western civilization might be food insecure seems bizarre. Food in grocery stores (and other outlets) is abundant, cheap, and delicious. Of all the things to worry about, food does not seem to be one of them![2] But the methods of industrial agriculture, which produce the vast share of this food, are beset with contradictions. A primary concern is reliance on declining reserves of fossil fuels, clean water, fertile topsoil, and materials for chemical fertilizers, pesticides, and herbicides. Industrial agriculture generates increasingly costly externalities including soil erosion and depletion, oceanic deadzones, and the destruction of bee and bat colonies (themselves critical to agriculture). This mode of production requires little human labor. It is managed with fossil fuel–based machines such as tractors, combines, airplanes (to spray pesticides and herbicides and distribute seeds), and other farm equipment. Most tellingly, industrial agriculture produces nearly 30% of Earth's greenhouse gases (Altieri, 2002; Vandermeer, 2011; Vermeulen et al., 2012; see Raghavan et al., 2016).

A further contradiction is organization of production around monocultures, an inherently fragile strategy for food production. With climate change, the dynamics of monoculture are even more likely to impact food security; a single pest or pathogen thriving under new conditions could decimate an entire crop. The easily replicated, generic methods of industrial agriculture have been around only since the end of World War II. Promoted by the US government and driven by the repurposing of wartime explosives factories (which turned to making fertilizer after the war), industrial agriculture has taken a decidedly unsustainable path (Vandermeer, 2011). In the spirit of collapse informatics, which seeks to plan ahead for a future of scarcity in the abundant present (Tomlinson et al., 2013), computational agroecology has been proposed as a way forward (Raghavan et al., 2016). Informed by the political economy of industrial agriculture, computational agroecology aims to turn industrial agriculture on its head by substituting knowledge and labor for expensive, damaging physical inputs. The approach promotes small- to medium-scale agriculture for all – planting food everywhere from rooftops and backyards to medium-sized farms. Crucially, agroecology creates *true biological ecosystems*, utilizing techniques from the science of agroecology. Starting in the 1980s, agroecology (a branch of the academic study of ecology) has been working out how to develop agricultural ecosystems with the same level of biodiversity as natural ecosystems (Vandermeer, 2011).

The techniques of computational agroecology share sensibilities with those of permaculture (Mollison, 1999), though they are more scientifically grounded. The techniques are consistent with some aspects of organic agriculture, though organic farmers often practice monoculture. Computational agroecology seeks to make agroecological design and management available to anyone through tools geared to localized conditions and resources. (Permaculture has attempted to diffuse its techniques through workshops, which has not scaled due to lack of experts who can serve as teachers). Computational agroecology is based on the idea of a sustainable political economy of accessible, scientifically managed food production of permanent crops such as fruit and nut trees, berry bushes, herbs, and perennial vegetables such as artichokes and asparagus, mixed with annual vegetables and flowers (see Raghavan et al., 2016). Computational agroecology can take advantage of the wealth of knowledge not only in scientific databases, but in communities of amateur gardeners and small farmers and growers, as well as knowledge of indigenous populations and their practices (see McCune, 2017).

Computational agroecology has strong resonance with the global social movement for *food sovereignty*, which is

> defined as the right of peoples and nations to create and maintain their own food systems. [Food sovereignty] has been at the heart of civil society protests against the free trade model imposed by international institutions such as the World Bank, the World Trade Organization, and the International Monetary Fund . . . Food sovereignty means a fundamental emphasis on local and domestic food production, based on land access for small farmers and ecological production practices. . . .
>
> (McCune, 2017)

Local control of food production, combined with intelligent management based on the science of agroecology, can leverage computational tools in an approach that aims for 'high-tech' *design and management* paired with accessible, cheap 'low-tech' *production techniques* such as simple machinery and the use of human labor. Information is the most potent ingredient in this mix. Information can be replicated and distributed digitally at near-zero marginal cost, allowing for wide adoption of practices underwriting food sovereignty, compared to the expensive inputs of industrial agriculture which are accessible only to those with money and extractive capacity. Capitalism drives toward the distribution of consumption, while hoarding production. Proposals for food sovereignty, computational agroecology, and other commons-based programs demand that productive capacity be released to all, for true security.

2. The second strategy for making inroads into the difficult problems of political economy and sustainability is to *use scientific logic to persuade and educate*. Education and persuasion are necessary, but we must remember that loud public alarms about ruinous environmental decline have been sounded since the 1970s (such as the *Limits to Growth* report), and, overall, things have gotten worse. We should keep pushing the logic and the science, but more is needed. We may need to change the ways in which we educate people (Ericksson and Pargman this volume). We do not believe logic and science should be abandoned solely in favor of 'stories' and the like. In the end, the stories must be informed by the science or they may turn out to be the wrong stories (just as Hollywood's narratives of apocalypse are rightfully grasped by audiences as mere entertainment). We are aware that people are weary of graphs and charts, and thus we hearken to efforts such as those of Ericksson and Pargman who seek to change the educational experience – a daunting but necessary task. Ericksson and Pargman note that a challenge is how to persuade people in ways that are not intellectually perplexing, emotionally overwhelming, or inept in practical ways.

In *Merchants of Doubt*, the influential book on denials of global warming, the authors discuss how solid scientific work has been rendered moot in most policy discussions, and, for that matter, in the public imagination. One line of their argument is "what matters in science is not what matters in politics", which leads them to suggest that, at least in the case of one study, "The devil was not in the details. It was in the main story" (Oreskes & Conway, 2010, pp. 172–73). The main story in this case was basically: CO_2 is a greenhouse gas, and it traps heat, so if you increase CO_2 Earth warms up. All the other details – that clouds, winds, and ocean circulation make a difference – were downplayed in the story because they essentially constituted "second-order effects – things that make a difference in the second decimal place, but not the first" (ibid.).

If we were to follow the same approach in talking about sustainable computing, we should ask, what is the main story? And what constitutes the first-, second-, and perhaps third-order effects? To answer these questions, we need to fall back on scientific findings, but we don't have to give the story in scientific detail. The issue is that, unlike the case of CO_2, which was shown to be a greenhouse gas as early as the mid-19th century by the Irish experimentalist John Tyndall, we might not yet have established theories for sustainable computing. What we do have, however, are some basic facts about the environmental hazards of unsustainable computing practices. But facts do not necessarily give birth to theories, so in order to tell a good story we might have to think about theory along the way – a theory comparable in simplicity to "CO_2 is a greenhouse gas." What would such a theory look like?

A good starting point, perhaps, would be to engage in a strategy of "reversal" (Agre, 1997; Ekbia & Sawhney, 2015), and bring attention back to the "material" basis of computing. To remind people, in other words, that, the cloud metaphor notwithstanding, computing involves an *essential* physical substrate. So the story is something like, "Computing needs a lot of *stuff* – both flesh and steel." Fortunately, this step has already been taken, as reflected in the growing literature on materiality. Unfortunately, the literature is often framed and cast in esoteric terms that are very foreign to the average person, starting with the term "materiality" itself. Fortunately, more accessible accounts of the physical basis of computing have become available, such as Andrew Blum's *The Tube: A Journey to the Center of the Internet*. However, narratives such as Blum's are overshadowed by dominant narratives of the cloud propagated by big business interests. We have to bring the physical narratives out of the shadow of the virtual ones. That is a core challenge for SHCI.

The second step is to dissect the "stuff" of computing into its key constituent components: steel *and* flesh. A political economy perspective might be useful in bringing both of these out. First, for "steel", this perspective can highlight the limited resources available on Earth, and the fights that are being fought in order to take control of those limited resources. Second, for "flesh", it can highlight the indispensible contribution of human labor to computing (Ekbia & Nardi, 2017).

A further step might focus on the implications of the second step – namely, that computing involves the exploitation of *both* physical and human resources. By highlighting both aspects, the story can come to life, allowing people to make a connection between their lived experience and the environmental impacts of computing. In parallel with the CO_2 story, then, the story of sustainable computing would be something like this:

> Computing needs a lot of stuff: steel and flesh, material and human. To increase computing, we need to exploit both Mother Nature and human beings, which implies the further depletion of natural resources.

Within this storyline, then, we can focus attention on the first-order effects of computing as far as sustainability is concerned. The reversal allows us to push second- and higher-order effects such as stimulation and entertainment to the margins, where they belong. The next question is, how can we implement and develop this as an educational strategy? To address this question, it would be useful to compare this with other approaches in sustainable computing such as the one developed by Ericksson and Pargman (this volume) and Pargman and Ericksson (2013).

Education in the trenches: an example

In an attempt to educate engineering students on issues of sustainability, Pargman and Ericksson (2013) introduced a three-dimensional "inspirational framework" that seeks to leverage "values" in order to help students understand their responsibilities as professionals *and* members of the society in developing "adaptation strategies", rather than the same old "mitigation strategies" of dealing with environmental risks. To that end, and based on what they describe as a "holistic view", they avoid a purely "fact-based" approach, and highlight instead the tensions that people face in thinking through these issues – namely, the tension between "delivering facts" and "discussing value", between "vanilla" and "doomsday" sustainability, and between technical/professional and personal/societal aspects of sustainability practices. On this last dimension, they raise the following important question:

> What is the relationship between, on the one hand, general issues pertaining to sustainability, for example personal behaviors, collective practices, societal infrastructure and global challenges, and on the other hand, the specific issues having to do with the topic of the course; ICT and media in relation to sustainability?
>
> (Pargman & Ericksson, 2013, p. 6).

The findings from the implantation of the above framework, based on student feedback, indicate mixed results, with some students finding inspiration and others feeling hopeless, guilt-ridden, and depressed, as captured in the vignettes reported in Ericksson and Pargman (this volume; see also Knowles & Ericksson, 2015).

The strategy proposed here from a political economy perspective maintains some of the elements of the above framework, but assembles and approaches them in a different fashion. First, in terms of "facts", the strategy incorporates them into a simple narrative that is easily communicable to broad audiences. Instead of counter-posing facts against values, it grounds them in a key aspect of computing – namely its materiality – that has been underplayed in dominant discourses. In this fashion, the strategy also grounds values *in* facts, rather than deriving them from a system that presumably transcends the real practice of computing. In the spirit of a pragmatist thinking, in other words, the *ought* of human behavior is rooted in the *is* of material practice (Dewey, 2005).

Second, in terms of the risk and severity of environmental collapse, the political economy perspective and the proposed narrative makes a clear connection to current capitalism and its urge for constant growth. Flesh and steel, the social and the natural, class exploitation and environmental destruction are brought together in a single framing, allowing people to see the connection through their lived experience. In doing so, responsibility is placed, first and foremost, with the socioeconomic system rather than with individuals, and feelings of guilt and hopelessness can be channeled toward action against the bigger system. The individual is not invited to relegate responsibility, rather to channel it toward bigger historical structures.

Lastly, in terms of the tension between our dual roles as technical professionals and ordinary citizens, the proposed strategy provides a balance by highlighting the ambivalent character of modern technology as being innovative and productive, on one hand, and undermining and destructive, on the other (Ekbia & Nardi, 2017). Rather than pushing individuals toward a forced choice between professional and social roles, the strategy allows them to perceive the former *within* the mandates of the latter.

All of this is not meant to suggest that the proposed strategy would provide a foolproof method of educating people. As with any other strategy, this one also involves its own tensions and challenges. It does, however, also have the potential for providing a holistic perspective, an engaged citizenship, and a balanced approach to individual and social responsibility.

3. A third strategy is to *devise future-facing technologies and practices* knowing that they will not be pervasively taken up any time soon, but that they will probably be useful some day. We need to frankly admit that there may be very little we can do *at scale* in the current moment because of the chokehold of the political economy. Clear and Comber (this volume) point to the magnitude and seriousness of the problems humanity faces. Critics of the current political economy underscore its tenacity and embeddedness (Burawoy, 1979; Gorz, 1985; Boltanski & Chiapello, 2005; Jackson, 2009; Harvey, 2010; Caffentzis, 2013; Klein, 2014; Piketty, 2014; Comor, 2015; Ekbia, 2016). The HCI community, and others, are working on technologies pertinent to these realities such as an Internet quine (Raghavan & Hasan, 2016), freecycling and barter (Knowles et al., 2013), timebanking (Bellotti et al., 2014), sustainable 3D printing (McDonald, 2016), strategies for limits-aware computing (Chen, 2016), commons for digital devices (Franquesa et al., 2016), approximate networking (Qadir et al., 2016), and, as discussed, computational agroecology (Raghavan et al., 2016).

Such technologies have two paths to widespread adoption, in our view. The first path runs within a scenario of collapse in which slow but relentless economic and environmental decline forces us to deal with planetary limits, and we draw from the creativity of those who saw the need to anticipate decline while still in a time of abundance. This agenda underlies collapse computing, which counsels us to prepare for a future of scarcity now, while it is easier with the resources we currently have (Tomlinson et al., 2013). The collapse scenario appears quite likely, and we believe that we should continue the lines of technological development being pioneered by the aforementioned scholars. These technologies emphasize self-reliance, drawing from a DIY ethic, rather than government or corporate policy or intervention (see Nardi, 2016).

The second path is to turn our attention to institutions, to take a step on the road to eliminating capitalism with its necessary ethos of growth,[3] by experimenting in ways that can begin to break down the toxic emphasis on social hierarchy and consumerism that fuels growth (see Gorz, 1985). Growth in the GDP sense of producing and consuming more things does not need to survive (in fact, it cannot because of a resource-constrained Earth), but some kind of institutional infrastructure beyond small localities does need to survive, or we will probably descend into societies shorn of the carefully constructed societal protections (civil rights, women's rights, LGBTQ rights, rights for the disabled, and so on) we have spent the last few hundred years accumulating, at no small human cost in time and energy, and even lives if we consider leaders such as Martin Luther King (see Jackson, 2009; Nardi, 2013).[4] It might seem odd to try to eliminate capitalism by using it as a testbed for its own demise, but we believe there is a strategy worth considering here. Perhaps it is premature to speak of dismantling capitalism, but if we take it to be a system that requires economic growth, then any long-range sustainability agendas must look to different ways of organizing economic activity.

Corporate purpose: an example

Institutional change seems remarkably difficult. At least with the self-reliant approaches, one can just start conducting the desired behavior. However, most of us don't want to become survivalists, so we are stuck with some version of institutional society. The trick is to alter what ails society, and find ways to turn our institutions to better account. These institutions are, after all, sleek, efficient miracles of productivity. We just need to use them for different societal objectives, such as cultural development, equitable healthcare, energy conservation, and economic equality.

We take inspiration from a remarkable paper called "The Problem of Corporate Purpose" that appeared in 2012 in the journal *Issues of Governance Studies*. Written by Lynn A. Stout, the Distinguished Professor of Corporate and Business Law at the Cornell School of Law, it calls into question the idea that the modern corporation exists only to increase shareholder value. Stout explains that what she calls the "shareholder value ideology" is, in fact, not supported by "the traditional rules of American corporate law . . . or by the bulk of the empirical evidence on what makes corporations and economies work". Among other things, this ideology "lure[s] companies into reckless and socially irresponsible behaviors." Stout observes that the shareholder value ideology has incurred "a daisy chain of costly corporate scandals and disasters, from massive frauds at Enron, HealthSouth, and Worldcom in the early 2000s, to the near-collapse of the financial sector in 2008, to the BP Gulf oil spill disaster in 2010, to the Walmart bribery scandal unfolding today". We might add that it has incurred increasing global economic inequality (Piketty, 2014) and increasingly unsustainable business practices and consumer behaviors, as detailed in these very pages, and in the voluminous literature the authors of this volume cite.

Stout explains that the *Dodge v. Ford* case in 1919, wherein the Michigan Supreme Court ruled that "a business corporation is organized and carried on primarily for the profit of the shareholders" is, in legal jargon, "mere dicta", i.e., it does *not* create binding precedent. We ourselves had always assumed that shareholders were a corporation's first and highest priority because of the prevalence of the ideology. Stout notes that legally, "[T]he business judgment rule holds that so long as a board of directors is not tainted by personal conflicts of interest and makes a reasonable effort to become informed, courts will not second-guess the board's decisions about what is best for the company – even when those decisions predictably reduce profits or share price." It is time to move beyond the outdated ideology of shareholder value, with its immense societal costs. The ideology is not legally constraining – corporations are free to pursue a wide variety of corporate purposes.

And herein lies the opportunity for SCHI. Stout shows that corporations are allowed to consider "long-term interests" – they do not need to slavishly produce quarterly profits that maximize short-term shareholder value. But in order for such a strategy to make sense, the corporation needs a compelling "corporate purpose" to mobilize its energies and measure its success. Stout does not say what those purposes should be, but she hints broadly, offering, for example, a story of the fisherman who maximizes his catch by using dynamite, a tactic that works for a while, but ends by spoiling the fishing for him and everyone else. Stout argues that the corporate imaginary of shareholders often reduces them "to their lowest possible common human denominator: shortsighted, opportunistic and untrustworthy, happy to impose external costs that reduce the value of other assets, and psychopathically indifferent to the welfare of other people, future generations, and the planet". Surely such shareholders are no longer a force for good in our world, and we should not encourage the ideology that produces them. We would not have believed such a change possible, but since distinguished mainstream scholars like Stout are advocating that we do exactly this, perhaps it *is* possible!

What about the corporations and their profits and their growth? Corporations will change in character, one way or the other (through collapse or through reinvention), and so will we all. As Jackson (2009) and others like Gorz have suggested, scenarios of zero growth (or even de-growth) indicate that we can all work less. We won't need as much money because we will consume less. Our quality of life should improve at least in terms of having time for activities other than paid work. Jackson cautions that merely advocating the destruction of our current institutions is not enough; we need to figure out a better way of doing things.

And this is where computing comes in. Reformulating a political economy is at bottom a computational task in that resources and labor must be reallocated. Raghavan and Pargman (2016) suggest that we "refactor" society in the computing sense, i.e., by reducing complexity while still retaining core functionality. This objective is an abstract research problem at this point, but just the sort of exciting challenge SCHI should take on. There would seem to be opportunities for crowdsourcing the data and opinions that would be needed to refactor society, bringing everyday citizens to the task of putting civilization on a sustainable course.

Just as the legal basis for corporations with long-term interests and societal purposes already exists, so do the computing models for refactoring. We have an astonishing intellectual basis to build on going forward. A key tenet of DIY is to use what is already at hand in creative ways, and we should do that, at the level of institutions.

Conclusion

The paradox of sustainability in the current political economy sometimes leads to moralizing that works against all three strategies we have proposed. Jackson, for example, spends a lot of time trying to convince us that we will be "happier" when we have less (2009). This is a red herring in our view. Many hold up the "local" as an ideal without considering the costs of trying to manage broad societal goals in each and every locality. Benjamin Franklin was probably correct in observing that, "There is no substitute for luxury." Can we question the wisdom of "I've been rich and I've been poor. Rich is better"? The goal of a future political economy, given Earth's limits, is not to dictate modes of happiness, but to provide basic freedoms, including freedom from destitution, within resource constraints. We believe that a pragmatic approach in which we keep our eyes on averting the precipice before us will be the most appealing and practical method for managing the challenge of sustainability.

We are in a good position to work toward this goal. An important function of the academy is to generate and distribute ideas. We do not make laws, lead political movements, or set corporate agendas. Yet our ideas percolate into the wider discourse in unpredictable ways. This leverage affords a chance to make a difference, even though it often feels like it's hard to know where to pin the tail on the donkey. But the opportunity is real, and we can labor to create ideas that are bold, just, and accessible to broad publics.

Notes

1 We are grateful to Professor Jen Mankoff for this point on choosing problems that benefit a small group.
2 We thought the same thing about clean water which turned out not to be true in, e.g., Flint, Michigan.

3 Tim Jackson's *Prosperity without Growth* (2009) draws from the work of American economist Jack Baumol to suggest that capitalism does not require growth and is merely about private property. We disagree, based on economics from Marx forward, including the authors we cite above, beginning with Burawoy. No less a sage than Benjamin Franklin observed that, "Money can beget money, and its offspring can beget more, and so on" (1748), emphasizing the underlying *mechanic of monetary growth* central to capitalism.
4 Jackson says, a little dramatically, that "a new barbarism lurks in the wings". We agree with the general point.

References

Agre, P. (1997). *Computation and Human Experience*. New York: Cambridge University Press.

Altieri, M. (2002). Agroecology: The science of natural resource management for poor farmers in marginal environments. *Agriculture, Ecosystems & Environment*, 93(1), 1–24.

Baumer, E., & Silberman, M. (2011). When the implication is not to design (technology). In *Proc CHI'11* (pp. 2271–2274).

Bellotti, V., Cambridge, S., Hoy, K., Shih, P., Handalian, L., Han, K., & Carroll, J. M. (2014). Towards community-centered support for peer-to-peer service exchange: Rethinking the timebanking metaphor. In *Proceedings of NordiCHI* (pp. 2975–2984).

Blevis, E. (2007). Sustainable interaction design: Invention & disposal, renewal & reuse. In *Proc CHI'07* (pp. 503–512).

Blyth, P., Mladenović, M., Nardi, B., Su, N., & Ekbia, H. (2015). Driving the self-driving vehicle: Expanding the technological design horizon. *Proc ISTAS* (International Symposium on Technology and Society Technical Expertise and Public Decisions). Dublin, Ireland.

Boltanski, L., & Chiapello, È. (2005). *The new spirit of capitalism*. New York: Verso.

Booth, D. (1998). *The environmental consequences of growth: Steady-state economics as an alternative to ecological decline*. London: Routledge.

Brown, D. (2015). *Undoing the demos*. Cambridge, MA: MIT Press.

Buechley, L., Rosner, D., Paulos, E., & Williams, A. (2009). DIY for CHI: Methods, communities, and values of reuse and customization. *CHI'09 Extended Abstracts*.

Burawoy, M. (1979). *Manufacturing consent: Changes in the labor process under monopoly capitalism*. Chicago: University of Chicago Press.

Caffentzis, G. (2013). *In letters of blood and fire*. Oakland: PM Press.

Cerratto Pargman, T., Pargman, D., & Nardi, B. (2016, May). The Internet at the eco-village: Performing sustainability in the twenty-first century. *First Monday*.

Chen, J. (2016). A strategy for limits-aware computing. *LIMITS '16*. Retrieved from http://limits2016.org/papers/a1-chen.pdf. AMC Workshop.

Comor, E. (2015). Revisiting Marx's value theory: A critical response to analyses of digital prosumption. *The Information Society*, 31(1), 13–19.

D'Allesandro, S., Luzzati, T., & Morroni, M. (2008). *GDP growth, consumption and investment composition: feasible transition paths towards energy sustainability*. Paper written for International Conference: DECROISSANCE ECONOMIQUE POUR LA SOUTENABILITE ECOLOGIQUE ET L'EQUITE SOCIALE.

Daly, H. (1991). *Steady state economics*. Washington, DC: Island Press.

Dewey, J. (2005). *Art as experience*. New York: Perigee. First published 1934.

Dillahunt, T. (2014). Toward a deeper understanding of sustainability within HCI. *CHI'14 Extended Abstracts*.

Dillahunt, T., & Mankoff, J. (2014). Understanding factors of successful engagement around energy consumption between and among households. In *Proc CSCW* (pp. 1246–1257).

DiSalvo, C., Sengers, P., & Brynjarsdottir, H. (2010). Mapping the landscape of sustainable HCI. In *Proc CHI '10*.

Ekbia, H. (2016). Digital inclusion and social exclusion: The political economy of value in a networked world. *The Information Society*.

Ekbia, H., & Nardi, B. (2014, June). Heteromation and its (dis)contents: The invisible division of labor between humans and machines. *First Monday*.

Ekbia, H., & Nardi, B. (2015, November/December). The political economy of computing: The elephant in the HCI room. *Interactions*.

Ekbia, H. R., & Nardi, B. (2017). *Heteromation and other stories of computing and capitalism*. Cambridge, MA: MIT Press.

Ekbia, H., Nardi, B., & Šabanović, S. (2015). On the margins of the machine: Heteromation and robotics. *Proc iConference*.

Ekbia, H. & Sawhney, H. (2014). Reason, resistance, and reversal: Metaphors of technology in design and law. *Culture, Theory, and Critique, 56*(2): 149–169. DOI: 10.1080/14735784.2014.904752

Franquesa, D., Navarro, L., & Bustamente, X. (2016). A circular commons for digital devices: Tools and services in eReuse.org. *LIMITS '16*. ACM Workshop. Retrieved from http://limits2016.org/papers/a3-franquesa.pdf

Friedman, B., & Nathan, L. (2010). Multi-lifespan information system design: A research initiative for the HCI community. *Proc CHI '10* (pp. 2243–2246).

Ganglbauer, E., Reitberger, W., & Fitzpatrick, G. (2013). An activist lens for sustainability: From changing individuals to changing the environment. *Proc 2013 International Conference on Persuasive Technology* (pp. 63–68).

Gorz, A. (1985). *Paths to paradise*. New York: South End Press.

Grudin, J. (1990). The computer reaches out: The historical continuity of interface design. *Proc CHI 1990* (pp. 261–268).

Gui, X., & Nardi, B. (2015). Sustainability begins in the street: A story of Transition Town Totnes. *Proc ICT4S*.

Håkansson, M., & Sengers, P. (2013). Beyond being Green: Simply living families and ICT. *Proc. CHI'13* (pp. 2725–2734).

Harvey, D. (2010). *The enigma of capital*. Oxford: Oxford University Press.

Hazas, M. (2015). Society pushes to go faster, but data binges carry environmental costs. *The Conversation*. Retrieved from https://theconversation.com/society-pushes-to-go-faster-but-data-binges-carry-environmental-costs-36672

Hazas, M., Morley, J., Bates, O., & Friday, A. (2016). Are there limits to growth in data traffic?: On time use, data generation and speed. *LIMITS '16*. DOI: http://dx.doi.org/10.1145/2926676.2926690

Hornborg, A. (2014). Why solar panels don't grow on trees. In K. Bradley & J. Hedrén (Eds.), *Green utopianism* (pp. 76–97). New York: Routledge.

Huang, E., & Truong, K. (2008). Breaking the disposable technology paradigm: opportunities for sustainable interaction design for mobile phones. *Proc. CHI '08*.

Huh, J., Nam, K., & Sharma, N. (2010). Finding the lost treasure: Understanding reuse of used computing devices. *Proc. CHI '10*.

Irani, L., & Silberman, S. (2013). Turkopticon: Interrupting worker invisibility in Amazon Mechanical Turk. In *Proc CHI'13* (pp. 16–21).

Jackson, S. (2014). Rethinking repair. In T. Gillespie et al. (Eds.), *Media technologies*. Cambridge, MA: MIT Press.

Jackson, T. (2009). *Prosperity without growth*. London: Routledge.

Kaptelinin, V. (2016). Making the case for an existential perspective in HCI research on mortality and death. In *Proceedings CHI 2016* (pp. 352–364).

Kim, S., & Paulos, E. (2011). Practices in the creative reuse of e-waste. *Proc. CHI '11*.

Kirman, B., Lawson, S., Linehan, J., & O'Hara, D. (2013). CHI and the future robot enslavement of humankind; a retrospective. *Proc CHI '13*.

Klein, N. (2014). *This changes everything: Capitalism vs the climate*. New York: Simon and Schuster.

Knowles, B., Blair, L., Hazas, M., & Walker, S. (2013). Exploring sustainability research in computing: Where we are and where we go next. *Proc Ubicomp '13* (pp. 305–314).

Knowles, B., & Ericksson, E. (2015, August). Deviant and guilt-ridden: Computing within psychological limits. *First Monday*.

Knowles, B. (this volume). Digital technology and political engagement.

Kuznetsov, S., & Paulos, E. (2010). Rise of the expert amateur: DIY projects, communities, and cultures. *Proc NordiCHI '10*.

Lomas, D., Kumar, A., Patel, K., Ching, D., Lakshmanan, M., Kam, M., et al. (2013). The power of play: Design lessons for increasing the lifespan of outdated computers. *Proc. CHI '13*.

Maestri, L., & Wakkary, R. (2011). Understanding repair as a creative process of everyday design. *Creativity and Cognition*, 81–90.

McCune, N. (2017). The long road: Rural youth, farming and agroecological formación in Central America. *Mind, Culture, and Activity*.

McDonald, S. (2016). 3D printing: A future collapse-compliant means of production. *LIMITS 2016*. ACM Workshop. Retrieved from http://limits2016.org/papers/a4-mcdonald.pdf.

Mollison, B. (1999). *Permaculture: A designer's manual*. Melbourne: Tagari Publications.

Morozov, E. (2011). *The net delusion: The dark side of internet freedom*. New York: Public Affairs.

Mota, C. (2011). The rise of personal fabrication. *Proc 8th ACM Conference on Creativity and Cognition*.

Nardi, B. (2013). The role of human computation in sustainability, or, social progress is made of fossil fuels. In *Handbook of human computation*. New York: Springer.

Nardi, B. (2016). Designing for the future – but which one? *Interactions*.

Nathan, L. (2008). Ecovillages, values, and information technology: Balancing sustainability with daily life in 21st century America. In *Proceedings of the ACM Conference on Human Factors in Computer Systems, Extended Abstracts* (pp. 3723–3728).

Oreskes, N., & Conway, E. M. (2010). Merchants of doubt: How a handful of scientists obscured the truth on issues from tobacco smoke to global warming. Bloomsbury Press: New York.

Packard, V. (1960). *The waste makers*. New York: David McKay.

Pargman, D., & Raghavan, B. (2014). Rethinking sustainability in computing: From Buzzword to non-negotiable limits. *Proc. NordiCHI '14*.

Patterson, D. (2015, August). Haitian resiliency: A case study in intermittent infrastructure. *First Monday*.

Pierce, J. (2012). Undesigning technology: Considering the negation of design by design. *Proc CHI '12* (pp. 957–966).

Piketty, T. (2014). *Capital in the twenty-first century*. Belknap Press.

Qadir. (2016). Taming limits with approximate networking. *LIMITS '16*. ACM Workshop. Retrieved from http://limits2016.org/papers/a9-qadir.pdf

Qadir, J., Sathiaseelan, A., Wang, L., & Crowcroft, J. (2016). Taming limits with approximate networking. *LIMITS '16*. doi>10.1145/2926676.2926678

Raghavan, R., & Hasan, S. (2016). Macroscopically sustainable networking: On Internet quines. *LIMITS '16*. ACM Workshop. Retrieved from http://limits2016.org/papers/a11-raghavan.pdf

Raghavan, B., Nardi, B., Lovell, S., Norton, J., Tomlinson, B., & Patterson, D. (2016). Computational agroecology: Sustainable food ecosystem design. *Proceedings CHI'2016*.

Raghavan, B., & Pargman, D. (2016). Refactoring Society: Systems Complexity in an Age of Limits. *LIMITS 2016*. ACM Workshop. Retrieved from http://dx.doi.org/10.1145/2926676.2926677

Remy, C., and Huang, E. (2015, August). Limits and sustainable interaction design: Obsolescence in a future of collapse and resource scarcity. *First Monday*.

Sabie, S., Salman, M., & Easterbrook, S. (2016). Situating Shelter Design and Provision in ICT Discourse for Scarce-resource Contexts. *LIMITS'16*. DOI: http://dx.doi.org/10.1145/2926676.2926686

Silberman, M. S. (2015). *Human-centered computing and the future of work: Lessons from Mechanical Turk and Turkopticon, 2008–2015*. PhD Thesis, University of California, Irvine.

Strauss, C., & Fuad-Luke, A. (2008). The slow design principles – a new interrogative and reflexive tool for design research and practice. Retrieved from www.slowlab.net/CtC_SlowDesignPrinciples.pdf

Suarez-Villa, L. (2015). *Corporate power, oligopolies, and the crisis of the state*. Albany: SUNY Press.

Tomlinson, B., Blevis, E., Nardi, B., Patterson, D. J., Silberman, M., & Pan, Y. (2013). Collapse informatics and practice: Theory, method, and design. *TOCHI*, *20*(4).

Toyama, K. (2015, August). Preliminary thoughts on a taxonomy of value for sustainable computing. *First Monday*.

Vandermeer, J. (2011). *The ecology of agroecosystems*. Jones & Bartlett.

Vermeulen, S., Campbell, B., & Ingram, J. (2012). Climate change and food systems. *Annual Review of Environment and Resources*, *37*(1), 195.

Weeks, K. (2011). *The problem with work: Feminism, Marxism, antiwork politics, and postwork imaginaries*. Durham, NC: Duke University Press.

Wong, J. (2009). Prepare for descent: Interaction design in our new future. *Proc CHI '09*, Sustainability Workshop position paper.

Woodruff, A., Hasbrouck, J., & Augustin, S. (2008). A bright green perspective on sustainable choices. *Proc CHI'08*.

6 Software engineering for sustainability

Tools for sustainability analysis

Birgit Penzenstadler and Colin C. Venters

Introduction

Sustainability is a central objective for systems development because, according to Rockström et al. (2009), humanity has exceeded the planet's boundaries for safe operation. Software systems are present in almost all aspects of our daily lives, therefore software systems can help to transition towards sustainability by acting as agents of change. If software engineers want to include sustainability as a new objective into their systems, it needs to be reflected in the requirements; therefore this may have significant impact on how software developers perform requirements engineering and systems design. However, software engineers lack practical guidelines on how to perform a sustainability analysis for a system while in discussion with customers. The key focus of this chapter is bridging the sustainability analysis gap in software engineering.

Human-computer interaction and software engineering

Human-computer interaction (HCI) is the study of how people interact with computers and to what extent computers are or are not developed for successful interaction with human beings. Software engineering (SE), in the context of sustainable HCI, means to develop software systems that take sustainability concerns into account from the way they impact social interaction to the different environmental impacts, as for example observed by Blevis (2007), Sengers (2011), Baumer and Silberman (2011), and Becker et al. (2016).

Brown (1997) compares SE and HCI methodologies and concludes that human-computer interface specialists are user-centered and software engineers are system-centered. Downton and Leedham (1991) integrated them most thoroughly by merging task analysis and user modeling (SE) with formal interface specifications and dialogue design tools (HCI), formal evaluation techniques (SE), and standards for documents (being) used to produce useful interactive software. This teaches us that while HCI needs to be user-centered and has own tasks, SE can still contribute a few tools to ease the integration of HCI design into the overall development process.

Software engineering and sustainability

In general, the relation between software engineering and sustainability has two main viewpoints. One is sustainable software, a rather technical perspective, which focuses on issues such as energy efficiency, maintainability, and extendibility. The other is software engineering for sustainability, which means software systems that positively contribute to an aspect of the sustainability of human (or other) life on this planet (Penzenstadler, 2013). As these two views are diametrically opposed, there is no common framework for sustainability in software engineering, even though various researchers have attempted to propose parts of what could become such a framework (Becker et al., 2015; Koziolek, 2011; Naumann et al., 2011; and Penzenstadler & Femmer, 2013).

Software Engineering for Sustainability (SE4S) denotes the concept of applying software engineering techniques to facilitate the refinement of higher-level sustainability concerns as defined in the domain into lower-level, technical requirements for the design and implementation of the system. The term was coined by Penzenstadler (2013), who investigated the relation between sustainability and software engineering (Penzenstadler et al., 2012; Penzenstadler et al., 2014).

Requirements engineering is the early part of the software development process, where the needs and wants of stakeholders are elicited, analyzed, and documented (Nuseibeh & Easterbrook, 2000). HCI design takes a similar approach, especially when it comes to integrating sustainability, because that requires a broader perspective than primarily the system (traditional SE perspective) or primarily the user (traditional HCI perspective). As Blevis (2007) states, "sustainability should be a central focus of interaction design".

While there have been attempts to understand how sustainability is perceived in the practice of software engineering and how sustainability can become an inherent part of software engineering practice (Chitchyan et al., 2016), consensus on what sustainability means in the field of computing is still emerging (Becker et al., 2015). However, it is suggested that rather than seeking broad conformity of definitions, the aim should be to clarify how the terms are used by different communities in order to have a shared understanding (Knowles et al., 2014). To increase the understanding and tangibility of this abstract concept, it is suggested that sustainability can only be evaluated with an exact reference to what is to be sustained? For whom? For how long? At what cost? (Tainter, 2003; Tainter & Taylor, 2014). In the field of software engineering, these questions mean the following:

- **What?** Requirements, by their nature, are a boundary-drawing exercise: what are the concerns for an application area, and how do we manage and prioritize those concerns? All else needs to be excluded from the analysis because otherwise the scope will be creeping infinitely. Requirements for sustainability, then, would seem to share similar challenges as "understanding and addressing sustainability" for a given context.
- **For whom?** The "for whom" is a particularly important negotiation, since reducing energy demand or environmental impacts is beneficial more widely,

but requirements are traditionally pitched at "the user". How would we capture requirements, which are sensitive to energy and environment? Do we involve facilities management? The local sustainability champion? Sustainability can be an inscrutable, ancillary concern (everyday life is still happening and might digress stakeholders), so who would be the environmental "stakeholder" we should consult when gathering requirements for "mundane practices", e.g., many of those in the home?

- **For how long?** In our current business models, planning horizons reach from few weeks for agile startups to years for large companies, but in general, there is an economically driven business plan. For better estimations of the long-term effects of our systems, it is necessary to include a long-term perspective of social and environmental perspectives in those business plans.
- **At what cost?** This question targets the trade-offs that we have to make for taking another objective into account. It goes hand in hand with the issue mentioned before – the long-term business plans.

This chapter offers two tools for how software engineering, in particular the early phase of requirements engineering, can help in solving the scoping (understanding what to sustain, for whom, for how long, and at what cost, as de-scribed in the previous paragraph) and more in depth-analysis of sustainability concerns:

- **Stakeholder Analysis:** What other stakeholders should be included in requirements elicitation? This is a more detailed exploration of the "for whom?" question that explores questions of responsibility and code of ethics.
- **Sustainability analysis:** What are the analysis boundaries of the system under consideration? How can we find those boundaries and choose the right system scope for development? What are the constraints implied by the system environment and by the sustainability goals? This is a more detailed exploration of the "what?", "for long?", and "at what cost?" questions.

The case study, which serves as running example for the discussion, is on resilient smart gardens with permaculture and sensors. The system under consideration is a project developed using an embedded board and permaculture principles (Norton et al., 2013). Finally, implications and next steps are discussed towards the end of the chapter.

Background and foundations

We explain our terminology and the most important concepts our work is based on. This includes sustainability concepts as well as the relationship between sustainability, software, and requirements engineering.

The sustainability concepts help to outline our understanding of sustainability for this chapter. The section on sustainability and software solidifies evidence why it is necessary to be concerned with sustainability in software development. The section on sustainability requirements sheds light on how sustainability comes into the discussion during requirements engineering and how they are being classified. Finally, the section on requirements engineering for sustainability provides an overview of our previous work on artifacts that support analyzing sustainability.

Sustainability concepts

This section introduces the orders of effect of information and communication technology on sustainability as well as the dimensions of sustainability we use in the remainder of this chapter.

On the basis of the simplest definition of sustainability as "the power to endure", we differentiate three orders of effect that are inspired by an original work of Hilty (2015): immediate, enabling, and structural.

- **Immediate effects** are the direct effects of the production, use, and disposal of software systems. This includes the immediate benefit of system features and the full lifecycle impacts, such as a lifecycle assessment (LCA) would include. An LCA evaluates the environmental impact of a product's life from the extraction of raw materials to its disposal or recycling.
- **Enabling effects** arise from a system's application over time. This includes not only opportunities to consume more (or fewer) resources but also other changes induced by system use.
- **Structural effects** represent persistent changes observable at the macro level. Structures emerge from the entirety of actions at the micro level and, in turn, influence these actions. Ongoing use of a new software system can lead to shifts in capital accumulation; drive changes in social norms, policies, and laws; and alter our relationship with the natural world.

These orders of effect are analyzed along the five dimensions of environmental, individual, social, economic, and technical sustainability, which were originally proposed in (Penzenstadler & Femmer, 2013). According to the most cited definition for sustainable development (UN World Commission on Environment and Development, 1987), sustainability is characterized by the three dimensions economic, social, and environmental. This characterization is extended with a fourth dimension human (or individual) by Goodland (2002), portraying the individual development of every human over his or her lifetime. When analyzing the sustainability of IT systems, these four dimensions apply, and an additional dimension, technical, supports better structuring of concerns with respect to software systems. Most concisely, we characterized the dimensions of sustainability in (Penzenstadler et al., 2013; Becker et al., 2015) as:

- **Individual sustainability** refers to maintaining human capital (e.g., health, education, skills, knowledge, leadership, and access to services).
- **Social sustainability** aims at preserving the societal communities in their solidarity and services.

- **Economic sustainability** aims at maintaining capital and added value.
- **Environmental sustainability** refers to improving human welfare by protecting the natural resources: water, land, air, minerals, and ecosystem services.
- **Technical sustainability** refers to longevity of systems and infrastructure and their adequate evolution with changing surrounding conditions.

Sustainability and software

Sustainability as a concept has emerged as an area of interest and topic of research within a number of subject areas in the field of computing including artificial intelligence (AI), high-performance computing (HPC), requirements engineering (RE), software engineering (SE), and human-computer interaction (HCI) as modern societies become increasingly dependent on complex software and software systems, which operate in evolving, distributed ecosystems (Geist & Lucas, 2009).

Software systems are now pervasive and underlie almost every aspect of modern societies from transportation, finance, education, retail, communication, governance, healthcare, and fitness, entertainment and leisure, defense and security, etc. (Deek et al., 2005). As a result, software systems themselves need to be sustainable. The problem is further compounded by change, as it is estimated that approximately 50–70% of the total lifecycle costs of a software system is spent on its maintenance and evolution (Ecklund et al., 1996). However, despite this increasing interest in the topic and the importance of software to modern societies, the concept of sustainability in relation to software is not well understood in the field of computing. Definitions of the concept range from a measure of time (Koziolek, 2011) to an emergent property (Katz et al., 2014).

As a result there is no consensus on how sustainability might be achieved with regards to the design of software artifacts and software systems, or how software sustainability might be quantified and measured. While this position is not unique to the field of computing, it is suggested that this lack of clarity about the concept of software sustainability will ultimately lead to ineffective and inefficient efforts to address the concept or result in its complete omission from the software system (Penzenstadler, 2013). As a result, the different fields interested in understanding and addressing sustainability face significant challenges in how to design, develop, test, maintain, and evolve these complex software system infrastructures. As such, one of the main challenges for the field of computing is to define the concept of sustainability and how it relates to and is understood by the different disciplines. From this we can derive how to design sustainable software from the existing bodies of knowledge, identify gaps in theory and best practice, all of which leads to forming an understanding of the basis of an educational curriculum which lays the foundation for training and educating the broad spectrum of computing students and advances the skills of existing practitioners to develop software that is sustainable.

Sustainability requirements

The term requirement has been in use in the field of software engineering since 1960 (Boehm, 2006). Similar to the term software sustainability or sustainable software, the term sustainability requirement has also emerged in the software and requirements engineering literature. This leads to a number of questions, including, is there such a thing as a sustainability requirement? If so, what are sustainability requirements? Are they the same as or are they different from other requirements? To provide a baseline for comparison we first consider the question, what are requirements from a requirements engineering perspective? Requirements in this field are generally classified into two broad categories: functional (FR) or non-functional requirements (NFR). In very simplistic terms, functional requirements describe what a system should do and non-functional requirements describe what a system should be. Software requirements express the needs and constraints placed on software that contributes to a real-world problem. At its most basic level, a software requirement is a property that the software system must exhibit. An essential property of all software requirements is that they must be verifiable.

While the term "sustainability requirement" has emerged in the literature, it is unclear how it is defined, applied, and understood, or how it is differentiated from how requirements are generally categorized. For example, Roher and Richardson (2013a) defined a sustainability requirement as "requirements that may be used to specify system behavior as well as to in influence the users behavior". In contrast, Huber et al. (2015) defined a sustainability requirement as "a requirement for a sustainable software system which concerns sustainability". In many cases, the term is not defined formally but is associated with one of the dimensions of sustainability such as environmental or economic (Roher & Richardson, 2013b).

A number of commentators suggest that sustainability as it relates to software should be classified as a first-class, non-functional requirement or software quality (Cabot et al., 2009; Mahaux et al., 2011; Koçak et al., 2015). In the field of requirements engineering, non-functional requirements or software quality attributes can be defined as the degree to which a system, component or process meets a stakeholder's needs or expectations (IEEE, 1990). This aligns with the Brundtland (1985) definition of sustainability addressing needs. While this view has been supported by a number of commentators, it has been made without explicit reference to the specific characteristics or qualities that sustainability as non-functional requirement or software quality would be composed of (Naumann, 2011; Calero et al., 2013; Seacord et al., 2003).

Similarly, a number of commentators have gone beyond this to define software sustainability as a composite, non-functional requirement, and identify the characteristics. For example, Venters et al. (2014a) argue that software sustainability is a measure of a systems extensibility, interoperability, maintainability, portability, reusability, scalability, and usability. At the very minimum, the genetic building blocks of [technically] sustainable software should address two core quality attributes: maintainability and extensibility, where the former is the effort required to locate and correct an error in operational software, and the latter is the software's ability to be extended and the level of effort required to implement the extension. A primary challenge is therefore how to demonstrate that the quality factors have been addressed in a quantifiable way. However, what quality attributes constitute sustainability as an NFR including their metrics and measures to demonstrate sustainability in software systems is an open research challenge (Venters, 2014b). In order to allow a better understanding of the meaning attributed to the term "sustainability requirements", Venters et al. (2017) investigated how current research constructs the notion of 'sustainability requirements' through published work? Their analysis suggested that the term "sustainability requirement" may be considered as a red herring in the sense that it is constructed in a way that suggests it is different in the way from how we understand requirements in general.

Requirements engineering for sustainability

Requirements engineering for sustainability (RE4S) can be supported by using an artifact-based approach developed in previous work by the first author (Penzenstadler, 2014). The approach was evaluated in several studies (Penzenstadler et al., 2015) and there is an online toolkit under development to support its application (Penzenstadler, 2016).

This chapter provides an updated and extended version of the approach in (Penzenstadler, 2014) as depicted in Figure 6.1. The approach starts with developing a stakeholder model and a goal model to get an overview of the needs and wants by the different parties involved in the project. Furthermore, the business context and the operational context need to be explored in a sustainability analysis to grasp the imposed constraints and understand the impacts the system may have. On that basis, a system vision is designed that has to be agreed upon by all stakeholders. One way of presenting a system vision that is easy to understand by technical as well as non-technical stakeholders is using rich pictures and illustrating the big picture of the system's main purpose(s) and how it may interact with the users and (potentially)

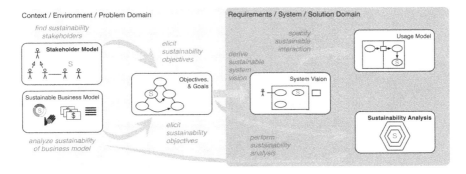

Figure 6.1 Requirements engineering for sustainability (with highlighted artifacts, stakeholder model, and sustainability analysis for the focus of this chapter)

with other systems. Subsequently, the development moves on to more detailed requirements elicitation, human-computer interaction design, and system design.

The stakeholder model and sustainability analysis we present below are crucial parts of this RE4S process and can significantly improve the support of sustainability as overall objective in development. The major new part in this updated version is the sustainability analysis inspired by previous work of Becker et al. (2016) where a sustainability analysis diagram was sketched out for a first time but never methodically described for its use in requirements engineering.

Stakeholder model: who are the advocates (and devil's advocates)?

This section explores what other stakeholders should be included in requirements elicitation. This is a more detailed exploration of the "for whom?" question mentioned in the introduction. It includes questions of responsibility and code of ethics, inspired by other fields like safety (where this challenge is solved) (Penzenstadler et al., 2014b).

Consider the situation of an analyst working on analyzing or improving the sustainability for a context (i.e., the concrete company or project under analysis). To ensure success of this undertaking, he needs to identify the involved stakeholders. To identify these stakeholders for sustainability there are four potential information sources that imply different, but simple approaches, which we describe in the following sections (see Figure 6.2):

1. Analyzing the dimensions to find responsible roles, and matching them top-down to the context.
2. Instantiating generic lists of sustainability stakeholders for the concrete context.
3. Inspecting the context, understanding which concrete roles are involved, and matching them bottom-up to the dimensions.
4. Iteratively analyzing and refining a generic sustainability goal model.

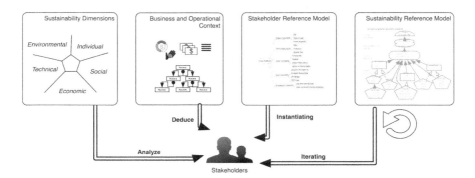

Figure 6.2 Ways for stakeholder identification

For applying the method, we expect that there is always one predominant (most suitable) information source that determines which of the approaches shall be used.

We document the stakeholders in a stakeholder model that allows listing and describing all stakeholders involved in a project. Stakeholders comprehend individuals, groups, or institutions having the responsibility for requirements and a major interest in the project. The Stakeholder Reference List in Table 6.1 is the result of an analysis of which stakeholders are impacted by or might be interested in a specific sustainability dimension. The left column lists the sustainability dimensions, the middle column the potentially affected stakeholders, and the right column the rationale of why that specific stakeholder is affected by that sustainability dimension. Thereby, some stakeholders show up multiple times, for example, legislation, as they affect multiple dimensions of the sustainability of a system and its environment. The table serves as checklist for ensuring that goals have been elicited and constraints have been collected from all stakeholders. Furthermore, the stakeholders have to validate the elicited requirements and sign off on them; therefore this list also serves to make sure all relevant decision-makers have been included.

The conclusion is: As the stakeholders are key for determining whether or not any objective is achieved, identifying the stakeholders for sustainability is crucial for successfully implementing and evaluating sustainability support in a given context (Penzenstadler et al., 2013). Furthermore, some of the stakeholders are relevant for the domain-dependent characterization of sustainability in a specific development context, for example activists and community representatives, while others are more relevant for the domain-independent implementation of a specific domain-dependent characterization of sustainability, for example, developers and maintenance personnel. Further details on all four approaches and the conducted case studies can be found in Penzenstadler et al. (2013).

Specifically in this book, Nardi points out that researchers and developers trying to address the sustainability challenge tend to overlook the fact that all countries' governments put economic growth first, and therefore the political economy is a perspective that we need to take into account in our models. Our stakeholder model takes care of this in its economic perspective.

Furthermore, Remy and Huang (this volume) look at stakeholders outside of user groups as well and present opportunities for engaging with target audiences that have yet to realize the benefits of SHCI research.

After all, who is the advocate for sustainability? We have identified a number of stakeholders for the dimensions of sustainability in different contexts. Roles that reoccurred across case studies are captured in the generic list in Table 6.1, which serves as a first reference checklist for further research and practice. We are positive that successfully identifying the stakeholders for sustainability will help ensure that this objective receives deserved attention.

Example: stakeholder model of resilient smart gardens

The system Resilient Smart Gardens with Permaculture and Sensors is a project developed using an Arduino board with a number of sensors for temperature

Table 6.1 Stakeholders affected by sustainability dimensions

Sustainability dimension	Stakeholder	Description/Rationale
Individual sustainability	User	The user is affected by the system in various ways. For example, users of online learning courses educate themselves through software.
	Developer	The developer is heavily involved in creating the system. Aspects like sustainable pace and growth of the developer must be considered.
	Employee representative	The mental and physical safety of individuals needs to be maintained. Employee representatives watch rights of employees involved.
	Legislation (individual rights)	Systems must respect the rights of their users. A legislation representative is a proxy for privacy and data protection laws.
Social sustainability	Legislation (state authority)	The state has a strong interest in understanding a system's influence on the society. Contrary to the individual rights legislation representative, the state authority representative speaks from the perspective of the state as a whole.
	Community representative	In addition to the state authority, other communities such as the local government (e.g., the mayor) or non-government clubs might be affected by a software system. A complete analysis must take their views into account.
	CRM	The customer relationship manager (CRM) is in charge of establishing long-term relationships with their customers and creating a positive image of the company.
	CSR manager	Some companies created the dedicated position of the corporate social responsibility (CSR) manager, who develops a company-specific vision of social responsibility.
Economic sustainability	CEO	The chief executive officer integrates sustainability goals into a company's vision.
	Project manager	It is very important to have the project manager agree in what ways the project should support sustainable aspects as he decides on prioritization with conflicting interests.

Sustainability dimension	Stakeholder	Description/Rationale
	Finance responsible	As sustainable software engineering often also affects the budget, many financial decisions have to be made to implement a sustainable software engineering model in a company.
Environmental sustainability	Legislation (state authority)	Environment protection laws are in place to ensure sustainability goals. These laws must be reflected in the requirements model for a system.
	CSR manager	The CSR manager is one of the responsible persons for environmental aspects and potential impacts on the business context.
	Activists /Lobbyists	Nature conservation activists and lobbyists (e.g., WWF, Greenpeace, BUND).
Technical sustainability	Admin	The administrator of a software system has a strong motivation for long-running, low-maintenance systems, making his work easier.
	Maintenance	The hardware maintenance is interested in a stable, long-term strategy for installation of hardware items.
	Customer	Users are interested in certain longevity of the systems they are using. This refers to user interface and required soft- and hardware.

and moisture. It was designed considering permaculture principles (Norton et al., 2013). These principles allow for sustainable long-term garden cultivation with maximum harvest based on the natural capacity of the soil enhanced by well-planned companion planting (Mollison, 1988; Holmgren, 2010).

The system vision is to connect a growing bed via sensors to a small embedded board such that we can measure moisture and temperature and log that data. This will enable determining the minimally feasible amount of watering, which is an environmentally sustainable measure in drought-prone Southern California. Future extensions include a connection to a plant database (to calculate typical needs of plants in the growing bed), to a database with geological information (to better predict the soil conditions), and a plant guild composer (Norton et al., 2013). This system will enable people with little background in gardening to successfully grow vegetables in the most sustainable and resource-conserving way.

Even though a prototype might not bring economic return on investment for private households because the investment compared to the water saving is rather low, a similar project did scale for a community garden as prototyped by the University of Central Florida in the project Connected Garden (Smith, 2013).

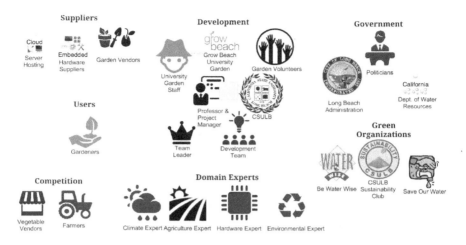

Figure 6.3 Stakeholder model of the resilient smart garden system

For the resilient smart garden system, 23 stakeholder groups were identified and then clustered into seven main types of stakeholders, see Figure 6.3.

1 Suppliers: Gardening supply vendors, IT hardware suppliers, and server hosts
2 Users: Gardener
3 Development: University garden staff, the Grow Beach initiative, garden volunteers, the project manager, the university, the team leader, and the development team
4 Government: the City of Long Beach administration, politicians, and the department of water resources
5 Competition: Vegetable vendors and farmers
6 Domain experts: Climate experts, agriculture experts, hardware experts, and environmental experts
7 Green organizations: Be Water Wise, CSULB Sustainability Club, and Save Our Water

Using the reference model, it was easy to identify stakeholders for each sustainability dimension, and having a complete list of interested parties made it easier to elicit a complete set of requirements that wouldn't miss any perspectives.

Sustainability analysis: what's the scope and relating impact?

What are the analysis boundaries of the system under consideration? How can we find those boundaries and choose the right system scope for development? We propose to analyze the potential effects that a system might have (Becker et al., 2016). Our approach includes the five dimensions of sustainability and three

Software engineering for sustainability 115

orders of effect, as introduced above in Sec. 2.1, as well as systems thinking concepts. By visualizing both the dimensions and the orders of effect in a radar chart, it is easier to keep an overview of the cross-relations than in any other type of diagram. In Figure 6.4, we provide an example analysis diagram for the resilient smart gardens.

The effects are on the radial axis and the dimensions are the sections of the radar chart. The system idea is the centerpiece. Now the requirements engineer or business analyst considers each dimensions and explores what immediate effects could occur from the system in use. Next, these effects are extrapolated to the enabling effects, and then to the structural effects. It is also possible that multiple enabling effects or structural effects are caused by an immediate effect, or that an enabling effect occurs where it is hard to pinpoint an underlying immediate effect. Furthermore, causal relations and impacts can occur across dimensions.

On that basis, a risk analysis can be performed to evaluate the likelihood of these impacts occurring under certain preconditions, and this is used to derive detailed requirements for achieving the desired effects (e.g., less environmental footprint) and for avoiding or reducing the not desired effects (e.g., costs due to technical debt).

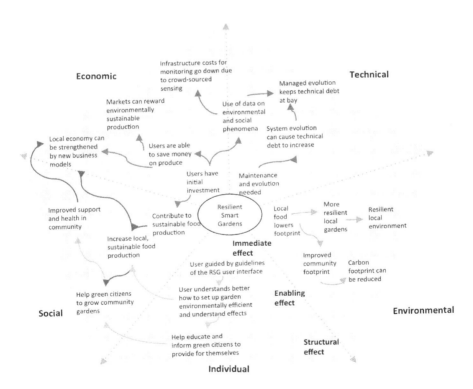

Figure 6.4 Sustainability analysis of the resilient smart garden system

Thus, the sustainability analysis diagram provides an easy-to-visualize technique to take a look at the bigger picture and long-term effects of a system in use in its larger environmental, social, economic, and technical context as well as specifically for the user (the individual dimension).

Example: sustainability analysis of resilient smart gardens

We argue that resilient smart gardens can become a contribution to sustainable development. Figure 6.4 provides an overview of the sustainability analysis we conducted for the resilient smart gardens. Starting from the individual dimension, as an immediate effect the user is guided by the guidelines of the resilient smart garden, as an enabling effect the users understand better how to set up their garden to be environmentally efficient and see the related environmental phenomena, and the structural effect can be that in the long run and assuming that the use of resilient smart gardens scales up to a significant percentage of the population, they help to educate and form green citizens.

In the social dimension, resilient smart gardens allow users to contribute to sustainable food production, which leads to the enabling effect of increased sustainable, local food production and, if the data is fed back into the central database, it enables the use of data about environmental and social phenomena. In the wider, structural effects, this can lead to improved support and health in communities (via food sharing) as well as crowd-created solutions for environmental and social problems related to food issues. The latter can lead to new business models being developed in the economic dimension.

For the economic dimension, an immediate effect are the costs associated with setting up a resilient smart garden, an enabling effect is that using sensors from a multitude of users for data collection on the environment reduces the sensing infrastructure costs, and in the long term that can lead to the dematerialization of monitoring (by outsourcing it to users).

For the technical dimension, there are the general maintainability and data collection issues, and system evolution may create new requirements, in the long run leading to large-scale IT infrastructures, in turn potentially concluding in high IT infrastructure costs in the economic dimension and increased CO_2 emissions in the environmental dimension.

In the environmental dimension, resilient smart gardens enable users to grow food in their own backyard in a sustainable and effective way, which long-term provides the ability to become more resilient as communities on a global scale.

All in all, the overall effects of resilient smart gardens can empower users to take care of growing their own food and sustaining themselves.

Conclusions

This chapter provided an exploration of how software engineering, more specifically requirements engineering, can be used to provide better for the needs of digital technology and sustainability and to confront the paradox that this often entails.

Taking responsibility for design

Sustainability is certainly not an exclusive requirements engineering concern, and sometimes the decision of whether to include sustainability as an objective for systems development may not even be up to the requirements engineer. Instead, when looking at the complete stakeholder model, it becomes clear that business analysts, marketers, and product portfolio managers should take an interest as well. It may not be up to the software designer to change their customers' motives, but software designers and interaction designers have as much responsibility as anyone else to learn, think, act, educate, and connect (Spinellis & Giannikas, 2012).

That said, we think that sustainability is a value that should be in the code of ethics for software engineers (and all other disciplines for that matter) in order to enable a more sustainable future. The authors have contributed a draft addition on that to the currently reworked version of the ACM code of ethics (http://ethics.acm.org/code-of-ethics/code-2018/). The authors are furthermore part of the working group around the Karlskrona Manifesto on Sustainability Design that promotes principles to help software engineers to consider sustainability as a major objective during design (Becker et al., 2015).

Future exploration: tie-in with interaction design research

There are two primary roads we see where HCI and software engineering research could connect better in the future and explore their interface:

1 Taking sustainable HCI research results and turning them into prototypes: There is a lot of great conceptual work from HCI research that remains at a stage of constructive and exciting philosophical discussion but has not made it into real-world implementations yet. Software engineering can help out here by collaborating and serving to develop the prototypes to allow HCI researchers to receive more data to evaluate the hypotheses underlying their concepts and strengthening their empirical evidence.
2 Making a business analysis including a focus on interaction way before developing a software system: Software engineers often fall for the trap of working in a manner that is too solution-oriented. Consequently, the developed software systems may not be the best fit for the actual needs they are trying to satisfy. Instead, the business process or service to be supported needs to be analyzed in depth with the help of interaction designers to find the right direction of support by a software system.

Engineering mindset versus limits to growth

As pointed out by Nardi (this volume) there is an inherent problem between our political world economy and the desire to transition human lifestyle towards sustainability.

From a designer's perspective, this is discussed by Baumer and Silberman (2011) in "When the implication is not to design?" – and that would be, for example, if we cannot achieve a significant simplification of reduction of overhead.

Similar concerns about solutionism are raised in Phoebe Senger's (2011) article on living on Change Islands for a long-term, design-ethnographic study.

Implications for education

The stakeholder model and the sustainability analysis are both taught within some of the software engineering sections by the first author. However, simply giving students these tools is not sufficient without first introducing them to the general concept of sustainability and the related issues around the contradictory promise of economic growth and planned obsolescence. This is also discussed in this book by Eriksson & Pargman (2017) in the chapter "On the inherent contradictions of teaching sustainability at a technical university".

References

Baumer, E. P., & Silberman, M. (2011, May). When the implication is not to design (technology). In *Proceedings of the SIGCHI Conference on Human Factors in Computing Systems* (pp. 2271–2274). New York: ACM.

Becker, C., Betz, S., Chitchyan, R., Duboc, L., Easterbrook, S. M., Penzenstadler, B., . . . Venters, C. C. (2016). Requirements: The key to sustainability. *IEEE Software, 33*(1), 56–65.

Becker, C., Chitchyan, R., Duboc, L., Easterbrook, S., Penzenstadler, B., Seyff, N., & Venters, C. C. (2015, May). Sustainability design and software: The karlskrona manifesto. In *Software Engineering (ICSE), 2015 IEEE/ACM 37th IEEE International Conference on* (Vol. 2, pp. 467–476). IEEE.

Blevis, E. (2007, April). Sustainable interaction design: invention & disposal, renewal & reuse. In *Proceedings of the SIGCHI conference on Human factors in computing systems* (pp. 503–512). New York: ACM.

Boehm, B. (2006, May). A view of 20th and 21st century software engineering. In *Proceedings of the 28th International Conference on Software Engineering* (pp. 12–29). New York: ACM.

Brown, J. (1997). HCI and requirements engineering-exploring human-computer interaction and software engineering methodologies for the creation of interactive software. *SIGCHI Bulletin, 29*(1), 32–35.

Brundtland, G. H. (1985). World commission on environment and development. *Environmental Policy and Law, 14*(1), 26–30.

Cabot, J., Easterbrook, S., Horkoff, J., Lessard, L., Liaskos, S., & Mazón, J. N. (2009, May). Integrating sustainability in decision-making processes: A modelling strategy. In *Software Engineering-Companion Volume, 2009. ICSE-Companion 2009. 31st International Conference on* (pp. 207–210). IEEE.

Calero, C., Bertoa, M. F., & Moraga, M. Á. (2013, July). Sustainability and Quality: Icing on the Cake. *RE4SuSy@ RE*.

Chitchyan, R., Becker, C., Betz, S., Duboc, L., Penzenstadler, B., Seyff, N., & Venters, C. (2016). Sustainability design in requirements engineering: state of practice. In

Proceedings of the 38th International Conference on Software Engineering Companion ICSE ('16) (pp. 533–542). ACM.
Deek, F. P., McHugh, J. A., & Eljabiri, O. M. (2005). *Strategic software engineering: An interdisciplinary approach.* CRC Press.
Downton, A., & Leedham, G. (1991). *Engineering the human-computer interface* (pp. 299–313). London.
Ecklund Jr, E. F., Delcambre, L. M., & Freiling, M. J. (1996, October). Change cases: Use cases that identify future requirements. *ACM SIGPLAN Notices, 31*(10), 342–358, ACM.
Eriksson, E., & Pargman, D. (2017). *On the inherent contradictions of teaching sustainability at a technical university.* London: Routledge.
Geist, A., & Lucas, R. (2009). Major computer science challenges at exascale. *The International Journal of High Performance Computing Applications, 23*(4), 427–436.
Goodland, R. (2002). Sustainability: Human, social, economic and environmental. *Encyclopedia of Global Environmental Change, 5,* 481–491.
Hilty, L. M., & Aebischer, B. (2015). ICT for sustainability: An emerging research field. In *ICT innovations for sustainability* (pp. 3–36). Springer.
Holmgren, D. (2010). *Permaculture: Principles & pathways beyond sustainability* (1st UK ed.). East Meon: Permanent Publications.
Huber, M. Z., Hilty, L. M., & Glinz, M. (2015). Uncovering sustainability requirements: An exploratory case study in canteens. In *RE4SuSy@ RE* (pp. 35–44).
IEEE. (1990, December). *Standard glossary of software engineering terminology.* IEEE Std 610.12–1990, pp. 1–84.
Katz, D. S., Choi, S. C. T., Lapp, H., Maheshwari, K., Löffler, F., Turk, M., . . . Swenson, S. (2014). Summary of the first workshop on sustainable software for science: Practice and experiences (WSSSPE1). arXiv preprint arXiv:1404.7414.
Knowles, B., Blair, L., & Walker, S. (2014). Toward a sustainability lexicon and pattern language? 1–3.
Koçak, S. A., Alptekin, G. I., & Bener, A. B. (2015). Integrating environmental sustainability in software product quality. In *RE4SuSy@ RE* (pp. 17–24).
Koziolek, H. (2011, June). Sustainability evaluation of software architectures: a systematic review. In *Proceedings of the joint ACM SIGSOFT conference – QoSA and ACM SIGSOFT symposium – ISARCS on Quality of software architectures – QoSA and architecting critical systems – ISARCS* (pp. 3–12). New York: ACM.
Mahaux, M., Heymans, P., & Saval, G. (2011, March). Discovering sustainability requirements: An experience report. In *International Working Conference on Requirements Engineering: Foundation for Software Quality* (pp. 19–33). Berlin, Heidelberg: Springer.
Mollison, B. (1988). *Permaculture: A designer's manual.* Melbourne: Tagari Publications.
Naumann, S., Dick, M., Kern, E., & Johann, T. (2011). The greensoft model: A reference model for green and sustainable software and its engineering. *Sustainable Computing: Informatics and Systems, 1*(4), 294–304.
Norton, J., Stringfellow, A. J., LaViola Jr, J. J., Penzenstadler, B., & Tomlinson, B. (2013). Plant guild composer: A software system for sustainability. In *RE4SuSy@ RE*.
Nuseibeh, B., & Easterbrook, S. (2000). Requirements engineering: A roadmap. In *Proceedings of the Conference on the Future of Software Engineering* (pp. 35–46). New York: ACM.
Penzenstadler, B. (2013, March). Towards a definition of sustainability in and for software engineering. In *Proceedings of the 28th Annual ACM Symposium on Applied Computing* (pp. 1183–1185). New York: ACM.

Penzenstadler, B. (2014). Infusing green: Requirements engineering for green in and through software systems. In *RE4SuSy@ RE* (pp. 44–53).

Penzenstadler, B. (2016). *Software engineering for sustainability toolkit*. Retrieved from www.csulb.edu/~bpenzens/se4s, last accessed 2016–11–02.

Penzenstadler, B., Bauer, V., Calero, C., & Franch, X. (2012). Sustainability in software engineering: A systematic literature review. In *Evaluation & Assessment in Software Engineering (EASE 2012), 16th International Conference on* (pp. 32–41). IET.

Penzenstadler, B., & Femmer, H. (2013). A generic model for sustainability with process- and product-specific instances. In *First Intl. Workshop on Green in Software Engineering and Green by Software Engineering*.

Penzenstadler, B., Femmer, H., & Richardson, D. (2013). Who is the advocate? Stakeholders for sustainability. In *2nd International Workshop on Green and Sustainable Software (GREENS, at ICSE)*.

Penzenstadler, B., Mehrabi, J., & Richardson, D. (2015). Supporting physicians by RE4S – evaluating requirements engineering for sustainability in the medical domain. In *Proceedings of the 4th International Workshop on Green and Sustainable Software at ICSE*.

Penzenstadler, B., Raturi, A., Richardson, D., Calero, C., Femmer, H., and Franch, X. (2014a). Systematic mapping study on software engineering for sustainability (se4s). In *Proceedings of the 18th International Conference on Evaluation and Assessment in Software Engineering* (p. 14). New York: ACM.

Penzenstadler, B., Raturi, A., Richardson, D., & Tomlinson, B. (2014b). Safety, security, now sustainability: The non-functional requirement of the 21st century. *IEEE Software Special Issue on Green Software*.

Remy, C., & Huang, E. M. (2017). *Communicating SHCI research to practitioners and stakeholders*. London: Routledge.

Rockström, J., Steffen, W., Noone, K., Persson, Å., Chapin, F. S., Lambin, E. F., . . . Nykvist, B. (2009). A safe operating space for humanity. *Nature*, *461*(7263), 472–475.

Roher, K., & Richardson, D. (2013a). A proposed recommender system for eliciting software sustainability requirements. In *User Evaluations for Soft-ware Engineering Researchers (USER), 2013 2nd International Workshop on* (pp. 16–19). IEEE.

Roher, K., & Richardson, D. (2013b). Sustainability requirement patterns. In *Requirements Patterns (RePa), 2013 IEEE Third International Work-shop on* (pp. 8–11). IEEE.

Seacord, R. C., Elm, J., Goethert, W., Lewis, G. A., Plakosh, D., Robert, J., . . . Lindvall, M. (2003, September). Measuring software sustainability. In *Software Maintenance, 2003. ICSM 2003. In Proceedings. International Conference on* (pp. 450–459). IEEE.

Sengers, P. (2011). What I learned on change islands: Reflections on it and pace of life. *Interactions*, *18*(2), 40–48.

Smith, E. (2013). *Connected garden*. University of Central Florida, E2i Creative Studio, Institute for Simulation and Training. Retrieved from http://e2i.ist.ucf.edu/project/8

Spinellis, D., & Giannikas, V. (2012). Organizational adoption of open source software. *Journal of Systems and Software*, *85*(3), 666–682.

Tainter, J. (2003). A framework for sustainability. *World Futures: The Journal of General Evolution*, *59*(3), 213–223.

Tainter, J., & Taylor, T. (2014). Complexity, problem-solving, sustainability and resilience. *Building Research & Information*, *42*(2), 168–181.

United Nations World Commission on Environment and Development. (1987). Re-port: Our common future. In *United Nations Conference on Environment and Development*.

Venters, C. C., Jay, C., Lau, L. M. S., Griffiths, M. K., Holmes, V., Ward, R. R., . . . Xu, J. (2014b). Software sustainability: The modern tower of Babel. *RE4SuSy 2014, In CEUR Workshop Proceedings* (Vol. 1216, pp. 7–12). CEUR.

Venters, C. C., Lau, L., Griffiths, M., Holmes, V., Ward, R., Jay, C., . . . Xu, J. (2014a). The blind men and the elephant: Towards an empirical evaluation framework for software sustainability. *Journal of Open Research Software*, 2(1).

Venters, C. C., Seyff, N., Becker, C., Betz, S., Chitchyan, R., Duboc, L., McIntyre, D., & Penzenstadler, B. (2017). *Characterising sustainability requirements: A new species, Red Herring, or just an odd fish?* 39th International Conference on Software Engineering ICSE 17, May 20–28, Buenos Aires, Argentina.

Response 2 Challenging the scope?

Enrico Costanza

The three chapters in Part 2 seem to challenge in different ways the scope of sustainable HCI. Nardi and Ekbia prompt us to consider the bigger picture framed through political economy, to better choose important problems to focus on, to communicate our work through scientific logic, and to consider technology that may become useful only in future sociopolitical scenarios. Joshi and Cerrato Pargman review definitions of the term "sustainable" within sustainable HCI and beyond. They then leverage that review to analyze user perceptions of a commercially available smartphone designed with the ambition of being sustainable. Penzenstadler and Venters put forward a methodology, comprising two conceptual tools, to take account of sustainability within software engineering, and especially requirements engineering.

I agree with several specific points made in the three chapters – just as one example, Nardi and Ekbia's suggestion to look at technologies that may become applicable only in future scenarios (*if* they materialize) resonates quite closely to some of my own work around autonomous systems mediating users' interaction with a future smart electricity grid (Costanza et al., 2014; Alan et al., 2016). However, considering the three chapters together, what I found striking is the rather wide range of levels of abstraction and the various scales of the projects analyzed by the authors.

On one end of the spectrum, Penzenstadler and Venters consider the "Resilient Smart Gardens" (RSG). It is an Arduino-based prototype, one that could be assembled by users themselves, leveraging easy-to-access and general-purpose prototyping components. The stakeholder model and the sustainability analysis of the RSG system are performed on the system in the form of a hardware prototype, rather than a possible future product which may be mass-produced commercially. Indeed, the main focus of the analysis of the RSG is on the software aspects of the system, true to the software engineering subject of the chapter.

In contrast, Joshi and Cerrato Pargman analyze the Fairphone (FP), a project that has long moved out of the prototyping stage, being now at its second generation of mass production, and having recently reached 100,000 products sold between the first and second product generations.[1] The analysis of the FP includes both the perceptions of users – who are notably also customers – and the ambition and communication strategies of the people working on the social enterprise behind the product (yet not quite the manufacturing and software engineering challenges, as I mention below).

On the opposite, far end of the abstraction spectrum, Nardi and Ekbia suggest that sustainable HCI needs to take into account political economy, and they put forward a small number of example ventures: from turning our attention to the *field* of computational agroecology to even "tak[ing] a step on the road to eliminating capitalism". While I am left with the curiosity of what Nardi and Ekbia would replace capitalism with (is this matter left open as an *exercise for the reader*?), I share their concern that limitless economic growth may well be at odds with sustainability.

I find the wide range of projects under analysis very intellectually stimulating; the three chapters prompted me to reflect not only about my own work as an HCI researcher, but also about my own everyday practices and consumption patterns as a citizen. Indeed, these chapters reminded me of the limits of HCI as a research discipline to tackle sustainability – is this the very paradox of sustainable HCI, then?

Is HCI sometimes too narrow as a lens to address sustainability? Without meaning to diminish the importance and propriety of the RSG analysis as an example of sustainable software engineering, do we need to go beyond software and consider the resources required to put in place the hardware (not only the embedded system and sensor, but even the servers) and communication infrastructure necessary for such a project, in order to assess its sustainability (e.g., Preist et al., 2016)?

Similarly, do we need to transcend HCI and take into account industrial, product, and software engineering to draw conclusions about the FP? The focus on user and stakeholder perceptions, as well as usability issues (e.g., the phone rebooting), makes the analysis presented by Joshi and Cerrato Pargman fall solidly within HCI. It seems natural, then, for such an analysis to leave out issues related to the manufacturing ecosystem around the smartphone, and its iterative product development, as these may require the lens of different disciplines. For example, the first version of the product, known as FP1, failed in some of its sustainability ambitions: on the FP1 the operating system could not be updated beyond a certain version, making the phone incompatible with current apps, and hence reducing its useful lifetime. The Fairphone team was quite open about the failure, publicly explaining how it happened.[2] Despite this failure, the social enterprise managed to form new partnerships and raise enough funding (at least in part through crowdfunding) to design a new phone model, the FP2, and start its mass production – perhaps a marketing achievement? A recent announcement[3] suggests that the FP2 succeeded where the FP1 failed: it is now possible to update the phone OS and thus extend the product lifetime.

I agree with Nardi and Ekbia's argument that problems such as water scarcity may be out of scope for "a typical 'user-centered' HCI approach", and that to address such a problem we probably need to think about agroecology, computational agroecology, and more in general rethink the political economy of food mass production (see also Petrini, 2007). I also do not doubt that within computational agroecology there may be opportunities and challenges for HCI, and that interactive technology may even have a role to play around possible future food networks (e.g., IoT and communication networks). However, I call into question whether it is always productive to extend the scope of sustainable HCI, rather than to acknowledge its limitations instead, and whether as researchers and as *citizens* we might need to think about sustainability and *act* beyond the limits of HCI as a research discipline.

Notes

1 www.fairphone.com/en/2016/05/26/100000-fairphone-owners/
2 www.fairphone.com/en/2014/12/09/our-approach-to-software-and-ongoing-support-for-the-first-fairphones/
3 www.fairphone.com/en/android-6-coming-to-fairphone-2/

References

Alan, A. T., Shann, M., Costanza, E., Ramchurn, S. D., & Seuken, S. (2016). It is too hot: An in-situ study of three designs for heating. In *Proceedings of the 34th Annual ACM Conference on Human Factors in Computing Systems* (pp. 5262–5273). New York: ACM.

Costanza, E., Fischer, J. E., Colley, J. A., Rodden, T., Ramchurn, S. D., Jennings, N. R. (2014). Doing the laundry with agents: A field trial of a future smart energy system in the home. In *Proceedings of the 32nd Annual ACM Conference on Human Factors in Computing Systems* (pp. 813–822). New York: ACM.

Petrini, C. (2007). *Slow food nation: Why our food should be good, clean, and fair*. New York: Rizzoli Ex Libris.

Preist, C., Schien, D., & Blevis, E. (2016). Understanding and mitigating the effects of device and cloud service design decisions on the environmental footprint of digital infrastructure. In *Proceedings of the 34th Annual ACM Conference on Human Factors in Computing Systems* (pp. 1324–1337). New York: ACM.

Photo Essay 4

Classroom exercise (2017). Students in a graduate interaction design class pose after responding to prompts to diagram concepts from an article by Elinor Ostrom, namely (i) the conventional theory of collective action, (ii) polycentric systems (of governance), (iii) can actions taken by multiple units cumulate to reduce the threat of climate change, and (iv) define: leakage, inconsistent policies, inadequate certification, gaming the system, and free riding. Reflection on Eriksson & Pargman: *On the inherent contradictions of teaching sustainability at a technical university*, as well as several other chapters that deal with public policy and interaction design. See: Elinor Ostrom. 2010. Polycentric systems for coping with collective action and global environmental change. *Global Environmental Change*, *20*(4), pp. 550–557.

Eli Blevis

Part 3
Ways to engage with others

7 Communicating SHCI research to practitioners and stakeholders

Christian Remy and Elaine M. Huang

Introduction

A decade has passed since the field of SHCI (sustainable human-computer interaction) emerged (Blevis, 2007; Mankoff et al., 2007), and numerous contributions to research have been made by the community. Alongside these considerable academic contributions, many members of the SHCI community have expanded their aspirations to include influencing broader environmental sustainability movements and communities. SHCI researchers have noted opportunities to communicate our work related to topics such as data centers (e.g., Preist et al., 2016), eco-feedback systems (e.g., Froehlich et al., 2010), or electronic waste (e.g., Remy et al., 2015). But HCI as a broad academic discipline has a long-standing "theory-practice gap" (Rogers, 2004; Roedl & Stolterman, 2013), which suggests that HCI researchers struggle to communicate their work to non-academic stakeholders and practitioners. This is the focus of our chapter. We raise the question of what the limitations of SHCI's current research are in bridging this gap, particularly in our approach to communicating research contributions to stakeholders outside of the SHCI community. The term "stakeholders" is used frequently in SHCI to refer to parties of interest outside of the field; they can be designers, manufacturers, policymakers, educators, or even the broad audience of "users" of a system. In this chapter, we aim to take a closer look at those groups, in particular focusing on communities SHCI has yet to properly engage with. Learning to communicate with those diverse stakeholders is an important challenge for SHCI researchers, especially for those who wish to bridge the theory-practice gap. Oftentimes, SHCI researchers aim not simply to advance technology, but to contribute to a sustainable future – a time-sensitive and complex issue that necessitates cooperation, coordination, and communication amongst many diverse stakeholders.

Therefore, the question for many SHCI researchers is: how can the field communicate its research to the various stakeholders outside the field such that it yields a noticeable real-world impact? Our approach to addressing this question comprises three steps: First, we revisit the theory-practice gap in the broader field of HCI and discuss how its lessons learned relate to SHCI. Second, we argue that the target audience for SHCI research is often not defined clearly enough and elaborate on why this is important and how this problem can be approached. Third, we highlight how we believe SHCI research needs to be framed in order to maximize impact on real-world practice. We hope that this discussion highlights opportunities for engaging with target audiences that have yet to realize the benefits of SHCI research for their practice, and sheds light on new avenues for research to improve the way we communicate our knowledge to those communities.

Revisiting the theory-practice gap

The term "theory-practice gap" refers to the well-known phenomenon in HCI that insights from research rarely make their way to the relevant practice beyond the research context (Rogers, 2004). A recent study highlighted that the issue is still prevalent (Roedl & Stolterman, 2013) and was also observed in SHCI research (Remy et al., 2015). It has been noted by design researchers (Roedl & Stolterman, 2013), in work related to interactive tabletops (Benko et al., 2009), and in theories for improving the design of user interfaces (Sutcliffe, 2000). However, there are a few success stories to be found that hint at potential solutions in bridging the gap. For example, with the Nest thermostat[1] a commercially successful solution gained a lot of attention in popular media in recent years (although it has not come out of research or have a direct connection, its design resembles findings from SHCI research). While its features focus on financial savings through heating optimization and the success in terms of permanent energy saving seems to have room for improvement (Yang et al., 2014), its design echoes a variety of conclusions the eco-feedback debate in SHCI has drawn over several years (e.g., Froehlich et al., 2010; Strengers, 2011; Pierce & Paulos, 2012).

Another increasingly popular, although not yet as widespread, design that aligns with SHCI findings is modular phones, such as the Fairphone,[2] Phonebloks,[3] or Project Ara.[4] The concept of modular devices has been brought up in SHCI frequently (e.g., Woolley, 2003; Blevis, 2007; Blevis et al., 2007; Huang & Truong, 2008; Dillahunt et al., 2010); whether any of those works have indeed influenced any of the aforementioned designs has not been conclusively determined. However, this highlights two additional problems of the theory-practice gap: First, oftentimes research is so theoretical and vague in its conclusions that its implementation in practice is hardly recognizable. Second, when research only offers theoretical knowledge and methods without any concrete and actionable guidelines or examples, its traces in practice are not only difficult to infer, but almost impossible to prove. This is exacerbated by the circumstance that for practice outside of research, in particular in terms of industrial development, but sometimes even in education or policymaking, there is no way of knowing whether solutions were directly inspired by SHCI research findings or not unless the involved parties are surveyed (which is oftentimes unfeasible and sometimes not possible due to certain restrictions).

Exceptions are when researchers take their work themselves into practice – such as in the case of SourceMap[5] (Bonanni et al., 2010) or Avaaj Otalo/Awaaz[6] (Patel et al., 2010). It is arguable to what extent those projects count as generalizable examples for bridging the theory-practice gap, since leaving research to bring one's own projects to life does not replace the need to communicate to existing stakeholders (e.g., large corporations or public institutions). The silver lining, however, clearly is that those examples show the value in research, if only such research has the chance to create an impact in a setting beyond its original research context. The amount of impact that research has is often more subject to external, uncontrollable factors (e.g., gain widespread attention in popular media, become viral on social networks, or reach critical mass of a large early adopter user base).

In the success stories about bridging the theory-practice gap, where research made it from HCI into real-world practice, we can identify one common thread: somewhere in the process the HCI knowledge was translated and made accessible to another, usually broader audience. This may have been through that popular media, which reported on research by translating the contributions from papers into articles digestible for a general audience. In the case of SourceMap and Avaaj Otalo, the researchers themselves took their contributions and translated them into practice, together with a team of practitioners (whom they taught). As we highlighted earlier, it is sometimes difficult to argue whether or not research had a direct impact on practice; if the research is purely theoretical, corresponding real-world practice cannot be linked to and the feedback channel is missing. However, in instances where research offers insights clearly applicable to practice and examples for overcoming the theory-practice gap are present, research needs to be translated such that theory and practice can meet and establish a connection. To get to this point, it is necessary to identify, understand, and engage the target audience and their practice in order to be able to successfully communicate research to them.

Identifying and engaging with the target audience

Oftentimes, research papers present concrete solutions to specific issues, but if their intention is to influence real-world practice they stay unnecessarily vague as to who those real-world practitioners are. Stakeholders, users, designers, policymakers, educators are broad terms for groups that encompass various levels of scale, expertise, and education, and this diversity poses fundamentally different challenges when being addressed. Phrasing SHCI research results using our traditional, theoretical, and abstract guidelines and principles can limit its accessibility beyond research communities and limit its real-world impact. We believe it is imperative to clearly define the target audience, become familiar with their specific requirements and needs, and engage with them as early as possible – potentially even throughout the entire research process by employing participatory design (DiSalvo et al., 2010; Silberman et al., 2014).

Acknowledging the complex network of stakeholders involved

A particularly striking example that we encountered in our own research was the sustainable design of consumer electronics, a topic very close to the concepts in sustainable interaction design (Blevis, 2007) and sustainability in design

(Mankoff et al., 2007). The problems we sought to address were electronic waste and obsolescence, for which SHCI research has produced a variety of solutions (Remy & Huang, 2015). The amount of e-waste is still growing (Baldé et al., 2015) and remains an unsolved and highly complex problem. What contributes to this complexity is the network of stakeholders involved. A simplistic perspective might identify two parties: the manufacturer, whose interests are mainly economical; and the consumer, who seeks pleasure and satisfaction (Woolley, 2003). However, a detailed inspection reveals that many more interest groups are involved and highly intertwined (Patrignani et al., 2011), including: "Chip Makers' Companies, Chip Manufacturing Workers, Cloud Computing Providers, Environment Advocacy Organizations, e-waste Destination Countries, Future Generations, ICT Vendors, ICT Users, Planet Earth and Policy Makers" (and this is "a simplified version of this network" (Patrignani et al., 2011)).

In this list, one can identify some of the stakeholders mentioned frequently in SHCI research addressing obsolescence, such as manufacturers and policymakers. Note the subtle distinctions though, such as between the company and its manufacturing workers, or environmental organizations and policymakers, which have different constraints, incentives, interests, and work practices. For example, if one were to attempt to influence public policymakers, the work of external organizations that fight for environmental sustainability should be taken into account, be it through work on the federal governmental level (e.g., lobbying) or more radical strategies utilizing public imagery (e.g., Greenpeace). Knowing those parties involved, their interactions, and the ways they influence each other is of critical importance to be able to successfully create an impact. One approach that researchers and organizations have pursued for many years is to address the consumer and hope for a feedback effect in which more environmentally conscious consumers force companies to take action and offer sustainable products. However, not only are other factors, such as price and aesthetics, sometimes more important to consumers (e.g., Hanks et al., 2008; Remy & Huang, 2012), this approach is also subject to "greenwashing" – companies exploiting the call for greener products by exaggerating their sustainable options. Approaching the problem head-on, e.g., by attempting to target the manufacturer rather than the user, can help mitigate those issues (e.g., Blevis, 2007; Khan, 2011).

While the stakeholders network is complex in the case of physical artifacts, such as our example of consumer electronics, it might even be more complex to model such a network in other problem areas that SHCI attempts to address. For example, when developing concepts that seek to raise awareness, change behavior, or promote knowledge, SHCI has often focused on the most visible stakeholders (i.e., the users) and neglected other parties (e.g., public policymakers or corporations). This issue is not new and has been brought up several times in seminal SHCI publications, with specific calls for researchers to shift focus from the individual to include other stakeholders or treat the common group of "users" differently (e.g., DiSalvo et al., 2010; Dourish, 2010; Brynjarsdottir et al., 2012; Silberman et al., 2014). Despite calls to change the practice of SHCI research, there are few clear signs of work that responds to those calls to action and fundamentally changes how researchers approach those problems – rather, research seems to shy away from this problem domain. We argue that instead SHCI researchers need to go one step further to identify the complex network of stakeholders. Who are additional potential stakeholders in those practices have not even been identified yet? When communicating sustainable information to consumers to raise awareness and facilitate behavior change, which parties are involved besides the researcher, the consumer, and the technological artifact?

We highlighted earlier that policymakers are influenced by external factors – public pressure, appointed advisors, lobbyists, and many more – and assuming those as a suitable target audience can be one potential lever to change the game and open up new avenues for research to create an impact. What if one of the shortcomings of eco-feedback technology and other research in SHCI that attempted to raise awareness and change behavior was that external stakeholders who influenced the decision-making and daily routines of users were not properly acknowledged? For example, aiming to change people's transportation behavior through eco-feedback technology could potentially be jeopardized by their social network or business contacts directly or indirectly influencing the decision-making process – potentially in the exact opposite way. We believe that it is critically important to make an effort to model the entire network of interest groups, at least to the extent possible, with all its facets to gain a holistic view of the problem domain and be able to identify new avenues for research. This increases the chance for SHCI research to successfully communicate their insights to practitioners outside of the community, by translating it into the target audience's language, as we have learned from previous theory-practice gap examples.

Engaging with the practitioners' practice

Knowing the stakeholders relevant to the problem domain is only the first step in communicating SHCI knowledge. Once we map the stakeholders, we should work to understand their unique practices and expertise. Doing so might allow us to identify opportunities to fuse SHCI research with the stakeholders' work processes, and bridge the theory-practice gap.

Going back to the example of designing sustainable consumer electronics to counter obsolescence, we identified in our research that the term "designer" was much too vague and did not sufficiently describe the work practices of those we sought to engage. Industrial designers, interaction designers, graphics designers, fashion designers, and even architects all fall under the umbrella term of designers and share commonalities in their practice, such as the iterative cycles when developing solutions. However, the detailed models of work practice of product design (Kruger & Cross, 2006), design engineering (Cross, 2008), architecture design (Lawson, 2006), and interaction design (e.g., Nielsen, 1994) show important differences. And even within one particular discipline other factors come into play, such as different levels of expertise (Kruger & Cross, 2006; Gonçalves et al., 2011) that highlight different problem-solving strategies (Kruger & Cross, 2006). Depending on those nuances, deciding on whether to translate SHCI design knowledge into textual or visual advice can have a significant impact on its use in practice (e.g., Muller, 1989; Goldschmidt & Smolkov, 2006; Goldschmidt & Sever, 2011). As highlighted in prior research, not only knowing those differences but translating research to be able to incorporate it into those designers' routines increases the chance to overcome the theory-practice gap (Remy et al., 2015).

While identifying the breadth of the potential target audience is the first step, investigating and understanding its everyday work practices are no less important task that should follow immediately. To communicate research to those stakeholders, previous examples we mentioned earlier highlighted that mediators translated the research into a language that was catered towards the target audience beyond the research community, as well as conveyed it in the most useful context. However, to be able to do this translation and be able to connect to the stakeholders, it is imperative to know their language and practices by engaging with them and getting to know their routines. For example, when news outlets cover research projects and translate them into online articles with the goal to go viral, they write them in a way that is not only understandable by a large audience (i.e., avoiding complex terminology), but frame them in a way that invites the reader and sparks their interest (i.e., "catchy headlines") and fits into their daily news consumption routines (i.e., short and simple rather than long and detailed). Therefore, in the next section we will discuss in detail how to transfer those insights and principles from translating research into our domain of SHCI knowledge.

How to frame our research to reach the relevant people?

When presenting SHCI research within our community, we can assume some degree of shared background, expertise, and expectations – thus, the framing of such research focuses less on the motivation but rather on the contribution. However, when communicating research to practitioners outside of SHCI, we should ask ourselves: How is SHCI research relevant to the target group's practices? Are they aware of existing SHCI research relevant to their practice? Are there opportunities for the target group to incorporate specific SHCI research insights into their real-world practice? Do they need an incentive to do so? These and other questions need to be addressed when framing SHCI research knowledge to be applied to practice outside of our field. HCI and SHCI research have already touched upon some of those questions (Dourish, 2006; DiSalvo et al., 2010; Silberman et al., 2014), and insights from previous attempts at addressing the theory-practice gap hint at possible solutions (e.g., Dalsgaard & Dindler, 2014; Remy & Huang, 2015). In the following, we discuss how researchers in SHCI could frame their work to make it accessible for practitioners and achieve successful knowledge transfer. We also acknowledge that this is sometimes at odds with typical research practice and highlight benefits as well as obstacles.

Translate SHCI knowledge

The first and most important step to make SHCI knowledge accessible for practitioners outside the field is to translate it into their domain's language. When researchers communicate their findings, such as in scientific presentations and publications, the language oftentimes contains genre-specific words and phrases that can pose a barrier for practitioners. For example, HCI has its roots in computer science and disciplines related to psychology; terminology that is considered to be basic jargon in those fields might lead to confusion or misunderstanding for people with other expertise. Moreover, the history, tools, and methods from those fields are the foundation of HCI research and are therefore critical for understanding research contributions; however, in aforementioned scientific presentations and papers they are often only briefly explained, if at all, since other researchers are familiar with those well-known foundations. Using genre-specific jargon and omitting details describing those foundations helps authors to keep their information more concise when communicating it to other researchers in the field, but can also make it difficult for practitioners to understand if they do not have this background knowledge.

Therefore, it is important to translate SHCI knowledge into the target audience's language to avoid such confusion or misunderstandings and make the research contributions more accessible. Translating can be understood literally in this case, as it describes the act of replacing words that are genre-specific terminology with terms familiar to the target audience, as well as describing practices that might be self-explanatory for domain experts. Oftentimes omitting details rather than describing them can help to avoid confusion ("less is more"), focusing on the results and the takeaways instead of elaborating on the methodology when it is not the key aspect the practitioner is interested in. An important point is to realize that for successful application of SHCI design knowledge, some practitioners might not necessarily need to understand how those contributions were conceived of, but they need to be able to incorporate it into their practice when most appropriate (Remy et al., 2015).

This leads to the second major point in translating SHCI knowledge: enabling practitioners to apply the research findings to their processes and incorporate it into their routines. Being able to understand SHCI knowledge bridges the first gap between researchers and practitioners, but it does not guarantee that the practitioners will be able to use them in their work practice. It is crucial for researchers to understand the target audience's work practice such that the findings can not only be translated, but also be appropriated to fit into their routines. For example, if the lessons learned from a research project are concrete guidelines for the choice of material in designed products, this might be more relevant to later stages in the design process (Kruger & Cross, 2006). Other guidelines and requirements, such as ISO standards or material properties as stated in the design brief, that compete with the SHCI design principles might come into play. In this case, the guidelines should be in a similar format and scope to other information provided at this stage. On the contrary, if SHCI design principles propose more foundational ideas that regard the overall concept of a designed object, applying those as early as possible in the process might be most beneficial (Khan, 2011).

Undeniably, this requires additional effort on the researcher's part: first and foremost, learning about the target audience's practice (e.g., from domain-specific literature or through exploratory studies such as interviews or surveys) – but also testing the translation of SHCI knowledge (e.g., recruiting practitioners and conducting a test run in a setting as close as possible to real-world practice). In the following, we discuss the opportunities that arise from this additional effort, but also how this knowledge transfer relates to existing research practice.

Opportunities and limitations

Translating scientific knowledge to make it more accessible to a broader audience is similar to the aforementioned example of news stories reporting on scientific findings: the headline and conclusion often boil research down to a single sentence, and fail to confront the reader with the complexity of research. While the skill set required is similar – being able to understand both the research as well as the receiving end and having the ability to translate knowledge from one to the other – there is an important difference when bridging the theory-practice gap. When communicating SHCI knowledge to practitioners, the target audience is known and quite narrow, unlike in the example of news about scientific contributions, which are meant to reach as many people as possible. This allows for a much more specific translation of knowledge; it is not just simplifying research contributions into shallow principles but adapting them to a different environment.

Several opportunities arise which researchers can seize in order to maximize benefit for their own research as well as the entire research field. First, engaging with the target audience to understand their practices and understand how one's research fits into it allows for a new perspective, potentially uncovering new directions for a specific project. Some of the contributions will prove to be more useful than others, and those pointers can be valuable for identifying which strains of research are most promising for future work. Second, the ability to communicate one's research to a different discipline and practitioners outside of the well-known research domain can prove to be a valuable skill in the future, for example, when negotiating funding for follow-up projects or attempting to gain more publicity for a research lab or entire conference. Third, bridging the theory-practice gap yourself rather than waiting for practitioners to pick up your findings and apply them yourself creates a direct feedback channel between research and practice. As mentioned in the beginning, practitioners take note of research and apply the findings in their daily processes, but oftentimes there is no feedback to the researchers once this knowledge transfer has occurred, in particular in the case of failure. Fourth, establishing connections to practitioners who can apply your research offers potential for collaboration and future research (e.g., by engaging in public policy projects (Thomas, this text) or teaching SHCI knowledge (Eriksson & Pargman, this text)).

Those opportunities come at the cost of additional effort; studying a different discipline and understanding their language, their work practices, and their daily routines takes time and resources, and so does translating SHCI design knowledge as well as testing the translation. This is oftentimes at odds with existing pressure of researchers to meet deadlines and publish at conferences and in journals – and communicating SHCI knowledge to practitioners is not a practice that can replace those activities. Rather, it can extend existing research practice, bring more exposure to a researcher's specific project as well as potentially the entire field, and therefore pay off in a different form, making up for the invested additional effort. We believe that in areas where engaging with practitioners is as crucial as in ours, due to the time pressure of issues of climate change, the additional effort is not only worth it but is almost required to keep SHCI active and alive. Targeting practitioners is also just one potential way to gain more exposure; others include engaging with issues of politics and economy (Nardi and Ebkia, this text) or investing into participatory design (Davis and Gram-Hansen, this text).

Conclusion

In this chapter, we argue that current SHCI research needs to rethink the way it is communicating its knowledge to practitioners outside the field to create a real-world impact on environmental sustainability. In particular, we highlight what the limitations of the current output of SHCI research are, why the vague definition of target audiences can jeopardize our research efforts, and how we need to frame our work differently. We emphasize once more that this is not to replace current publication practice, but add new, additional ways of communication – specifically suited towards reaching out to practitioners and stakeholders.

Notes

1 https://en.wikipedia.org/wiki/Nest_Labs
2 www.fairphone.com/
3 http://phonebloks.com/
4 https://atap.google.com/ara/
5 www.sourcemap.com/
6 www.awaaz.de/

References

Baldé, K., Wang, F., Huisman, J., & Kuehr, R. (2015). *The global e-waste monitor – 2014*. Bonn, Germany: United Nations University, IAS – SCYCLE.

Benko, H., Morris, M. R., Brush, A. J. B., & Wilson, A. D. (2009). *Insights on interactive tabletops: A survey of researchers and developers* (No. MSR-TR-2009–22).

Blevis, E. (2007). Sustainable interaction design: Invention & disposal, renewal & reuse. In *Proceedings of the SIGCHI Conference on Human Factors in Computing Systems* (pp. 503–512). New York, NY: ACM.

Blevis, E., Makice, K., Odom, W., Roedl, D., Beck, C., Blevis, S., & Ashok, A. (2007). Luxury & new luxury, quality & equality. In *Proceedings of the 2007 Conference on Designing Pleasurable Products and Interfaces* (pp. 296–311). New York, NY: ACM.

Bonanni, L., Hockenberry, M., Zwarg, D., Csikszentmihalyi, C., & Ishii, H. (2010). Small business applications of Sourcemap: A web tool for sustainable design and supply chain transparency. In *Proceedings of the SIGCHI Conference on Human Factors in Computing Systems* (pp. 937–946). New York, NY: ACM.

Brynjarsdottir, H., Håkansson, M., Pierce, J., Baumer, E., DiSalvo, C., & Sengers, P. (2012). Sustainably unpersuaded: How persuasion narrows our vision of sustainability.

In *Proceedings of the 2012 ACM Annual Conference on Human Factors in Computing Systems* (pp. 947–956). New York, NY: ACM.

Cross, N. (2008). *Engineering design methods: Strategies for product design*. Chichester, UK; Hoboken, NJ: Wiley.

Dalsgaard, P., & Dindler, C. (2014). Between theory and practice: Bridging concepts in HCI research. In *Proceedings of the SIGCHI Conference on Human Factors in Computing Systems* (pp. 1635–1644). New York, NY: ACM.

Dillahunt, T., Mankoff, J., & Forlizzi, J. (2010). A proposed framework for assessing environmental sustainability in the HCI community. In *Examining Appropriation, Re-Use, and Maintenance of Sustainability Workshop at CHI 2010*.

DiSalvo, C., Sengers, P., & Brynjarsdóttir, H. (2010). Mapping the landscape of sustainable HCI. In *Proceedings of the SIGCHI Conference on Human Factors in Computing Systems* (pp. 1975–1984). New York, NY: ACM.

Dourish, P. (2006). Implications for design. In *Proceedings of the SIGCHI Conference on Human Factors in Computing Systems* (pp. 541–550). New York, NY: ACM.

Dourish, P. (2010). HCI and environmental sustainability: The politics of design and the design of politics. In *Proceedings of the 8th ACM Conference on Designing Interactive Systems* (pp. 1–10). New York, NY: ACM.

Froehlich, J., Findlater, L., & Landay, J. (2010). The design of eco-feedback technology. In *Proceedings of the SIGCHI Conference on Human Factors in Computing Systems* (pp. 1999–2008). New York, NY: ACM.

Goldschmidt, G., & Sever, A. L. (2011). Inspiring design ideas with texts. *Design Studies*, *32*, 139–155.

Goldschmidt, G., & Smolkov, M. (2006). Variances in the impact of visual stimuli on design problem solving performance. *Design Studies*, *27*, 549–569.

Gonçalves, M. G., Cardoso, C., & Badke-Schaub, P. (2011). Around you: How designers get inspired. In *Proceedings of the 18th International Conference on Engineering Design*.

Hanks, K., Odom, W., Roedl, D., & Blevis, E. (2008). Sustainable millennials: Attitudes towards sustainability and the material effects of interactive technologies. In *Proceedings of the SIGCHI Conference on Human Factors in Computing Systems* (pp. 333–342). New York, NY: ACM.

Huang, E. M., & Truong, K. N. (2008). Breaking the disposable technology paradigm: Opportunities for sustainable interaction design for mobile phones. In *Proceedings of the SIGCHI Conference on Human Factors in Computing Systems* (pp. 323–332). New York, NY: ACM.

Khan, A. (2011). Swimming upstream in sustainable design. *Interactions*, *18*, 12–14.

Kruger, C., & Cross, N. (2006). Solution driven versus problem driven design: Strategies and outcomes. *Design Studies*, *27*, 527–548.

Lawson, B. (2006). *How designers think: The design process demystified*. Oxford; Burlington, MA: Elsevier/Architectural.

Mankoff, J. C., Blevis, E., Borning, A., Friedman, B., Fussell, S. R., Hasbrouck, J., ... Sengers, P. (2007). Environmental sustainability and interaction. In *CHI '07 Extended Abstracts on Human Factors in Computing Systems* (pp. 2121–2124). New York, NY: ACM.

Muller, W. (1989). Design discipline and the significance of visuo-spatial thinking. *Design Studies*, *10*, 12–23.

Nielsen, J. (1994). *Usability engineering*. San Francisco, CA: Morgan Kaufmann Publishers.

Patel, N., Chittamuru, D., Jain, A., Dave, P., & Parikh, T. S. (2010). Avaaj Otalo: A field study of an interactive voice forum for small farmers in rural India. In *Proceedings of the SIGCHI Conference on Human Factors in Computing Systems* (pp. 733–742). New York, NY: ACM.

Patrignani, N., Laaksoharju, M., & Kavathatzopoulos, I. (2011). Challenging the pursuit of Moore's law: ICT sustainability in the cloud computing era. *POLITEIA, XXVII.*

Pierce, J., & Paulos, E. (2012). Beyond energy monitors: Interaction, energy, and emerging energy systems. In *Proceedings of the SIGCHI Conference on Human Factors in Computing Systems* (pp. 665–674). New York, NY: ACM.

Preist, C., Schien, D., & Blevis, E. (2016). Understanding and mitigating the effects of device and cloud service design decisions on the environmental footprint of digital infrastructure. In *Proceedings of the SIGCHI Conference on Human Factors in Computing Systems* (pp. 1324–1337). New York, NY: ACM.

Remy, C., Gegenbauer, S., & Huang, E. M. (2015). Bridging the theory-practice gap: Lessons and challenges of applying the attachment framework for sustainable HCI design. In *Proceedings of the SIGCHI Conference on Human Factors in Computing Systems* (pp. 1305–1314). New York, NY: ACM.

Remy, C., & Huang, E. M. (2012). The complexity of information for sustainable choices. In *Simple, Sustainable Living Workshop at CHI 2012.*

Remy, C., & Huang, E. M. (2015). Addressing the obsolescence of end-user devices: Approaches from the field of sustainable HCI. In L. M. Hilty & B. Aebischer (Eds.), *ICT innovations for sustainability* (p. 474). Switzerland: Springer International Publishing.

Roedl, D. J., & Stolterman, E. (2013). Design research at CHI and its applicability to design practice. In *Proceedings of the SIGCHI Conference on Human Factors in Computing Systems* (pp. 1951–1954). New York, NY: ACM.

Rogers, Y. (2004). New theoretical approaches for human-computer interaction. *Annual Review of Information Science and Technology, 38,* 87–143.

Silberman, M. S., Nathan, L., Knowles, B., Bendor, R., Clear, A., Håkansson, M., . . . Mankoff, J. (2014). Next steps for sustainable HCI. *Interactions, 21,* 66–69.

Strengers, Y. A. A. (2011). Designing eco-feedback systems for everyday life. In *Proceedings of the SIGCHI Conference on Human Factors in Computing Systems* (pp. 2135–2144). New York, NY: ACM.

Sutcliffe, A. (2000). On the effective use and reuse of HCI knowledge. *ACM Transactions on Computer-Human Interaction, 7,* 197–221.

Woolley, M. (2003). Choreographing obsolescence – ecodesign: The pleasure/dissatisfaction cycle. In *Proceedings of the 2003 International Conference on Designing Pleasurable Products and Interfaces* (pp. 77–81). New York, NY: ACM.

Yang, R., Newman, M. W., & Forlizzi, J. (2014). Making sustainability sustainable: Challenges in the design of eco-interaction technologies. In *Proceedings of the SIGCHI Conference on Human Factors in Computing Systems* (pp. 823–832). New York, NY: ACM.

8 Negotiating and engaging with environmental public policy at different scales

Vanessa Thomas

Introduction

Much like computing, public policy is ubiquitous. It influences what we can do; where, when, and how we can pursue available services or activities; and who is involved along the way. The wide reach of public policy has inspired members of the HCI community to host public policy forums, write public policy papers, and call for increased engagement with the actors and governance tools that shape public policy (Lazar et al., 2016). These latter calls to engage with policy have extended to the environmental sphere (Goodman, 2009; Håkansson & Sengers, 2014, Knowles et al., 2014; Silberman et al., 2014) and have opened up several exciting questions for the HCI community to consider. These questions include: With which environmental public policies should we engage? How should we engage with those policies? For what aims, and at what scales, should we engage with them? Few clear answers have been proposed to these questions, perhaps because few clear answers exist. Environmental public policy (EPP) is a broad domain to navigate, and opportunities for the HCI community to engage with EPPs might appear limited. However, many EPPs speak directly to issues addressed by, or of interest to, the HCI community, and a concrete discussion of the aforementioned questions seems timely.

This chapter explores some of the ways that public policy, human-computer interaction (HCI), and environmental sustainability intersect. Each of those domains have their own complex and contested histories, practices, terms, theorists, applications, politics, and methods, and this chapter merely offers an introductory discussion of how we might negotiate and engage with their overlapping territory. As part of that discussion, this chapter examines two environmental public policy domains: "waste electrical and electronic equipment" (WEEE) and "green public procurement" (GPP) policies. Public policy research communities have discussed these policy domains at length, and they relate directly to some of Blevis' original sustainable interaction design principles (Blevis, 2007). I introduce how these policies operate at different scales, and then discuss how they could inform the work of the HCI community, and how they could be informed by the work of the HCI community. By "different scales", I mean to address the 'domestic' (e.g., municipal or national) and 'international' (e.g., United Nations [UN] or African Union [AU]) policy scales identified by Nathan and Friedman (2010).

This chapter thus expands upon the description of public policy, framing of two EPPs (WEEE+GPPs), as well as the discussion around engagement, found in a CHI note (Thomas et al., 2017); readers may also be interested in the policy brief which the author prepared for Global Affairs Canada, supplied as supplemental material to that note, on the ACM Digital Library. This chapter begins by introducing public policy and its connections to HCI research. It then offers a preliminary discussion of how and why we might engage with WEEE and GPP policies at the domestic and international policy scales. The chapter closes by examining some of the ways we might engage with EPPs and some of the challenges we might encounter. Those challenges are not meant to deter us; there are many EPP issues and domains that are ripe for future exploration, and the HCI community is well positioned to engage with their socio-technical and environmental implications.

Public policy and HCI

At its most basic level, public policy can be thought of as the rules and objectives that governments set for themselves (Howlett & Ramesh, 2003, pp. 5–11). Governments draft and adopt public policies to guide their action and inaction on diverse areas of concern (Dye, 1978; Howlett & Ramesh, 2003, pp. 5–11), and these policies can dictate the types of programs and projects that governments fund, as well as the partnerships that they seek to establish. For example, Her Majesty's Government (i.e., the British government) has adopted an "open government" policy and now maintains a list of public policies on their website (HMSO, 2017). The topics covered by their policies range from accessible transport and housing for older and vulnerable people to energy demand reduction and biodiversity (HMSO, 2017). Each policy section on the gov.uk website includes documentation of relevant laws, case studies, impact assessments, statistics, public consultations, regulations, news stories, and international treaties (HMSO, 2017).

Policies play a unique and iterative role in government and society; they are often based on existing laws, regulations, strategies, and social norms, but they also influence future laws, regulations, strategies, and social norms (Kraft & Furlong, 2014, pp. 3–6). As a result, the process of understanding policies, or undertaking policy analysis, often relies on going beyond what a policy or official government document states, and "necessarily involves [analyzing] the complex array of state and societal actors involved in decision-making processes" (Howlett & Ramesh, 2003, p. 7). Many societal actors are involved with developing, setting, and influencing public policy, including legal experts, professional societies, interest groups, academics, private companies, and non-profit organizations (Lazar et al., 2016, pp. 71–74). Even other levels of government – such as city-level vs. national – and intergovernmental organizations (e.g., the European Union [EU], AU, UN) can influence public policy across the national and international policy scales, as implied by Nathan and Friedman's (2010) domestic and international scales.

Members of the HCI community have been amongst the societal actors engaged directly and indirectly with public policy. Some have been engaging with public policy for decades. In the 1980s and 1990s, several HCI researchers and practitioners worked to influence "accessibility" policies in the US (Lazar et al., 2016). They designed interfaces that would be more usable for people with disabilities, and worked to influence technical standards related to those interfaces (Lazar et al., 2016). Similar work on accessibility has continued to this day, with members of the HCI community examining and engaging with accessible designs, policies, and laws around the world (e.g., Danielsen et al., (2011) and Gulliksen et al. (2010)). Some members of the community have also engaged with human rights (Nathan & Friedman, 2010), urban development (Crivellaro et al., 2016; Vlachokyriakos et al., 2016), and arts/innovation (Fantauzzacoffin et al., 2012) policies, as well as a few international technical standards (Lund et al., 2012; Mirnig et al., 2015). Others have written magazine articles, some have worked with governmental organizations or been funded by government research bodies, and some have issued – as well as responded to – calls for activism related to complex social issues (Davis et al., 2012). These actions and endeavors have all directly or indirectly affected public policy.

Environmental public policies and HCI

Environmental public policy is a broad group of policies that attempt to address environmental issues, including energy production and consumption, non-renewable resource extraction, biodiversity, water and food security, and "green" economic growth (Elsevier B.V, 2017). These policies range in scope and scale from locally specified waste management policies to nationally set objectives for reducing greenhouse gas emissions and internationally negotiated methods for combating climate change. In this section, I describe two sets of environmental public policies – GPP and WEEE policies – and how they operate at different scales. I then explore what that means for the HCI community using the two-pronged framework outlined by Lazar et al. (2016) to address: how EPPs can influence the HCI community, and "how the HCI community can inform public policy by providing expertise, taking part in the development of policy, and researching the impact of various policies related to HCI" (p. 74).

Waste Electrical and Electronic Equipment (WEEE) public policies

Throughout the 1990s and early 2000s, countless news stories and research projects exposed the international environmental and human health damage caused by the export of waste electrical and electronic equipment. Many businesses in the Global North had been shipping their hazardous WEEE to countries in the Global South, where the highly toxic materials were being dumped in open pits or processed in high-risk informal waste sites (Lepawsky, 2014; Widmer et al., 2005; SBC, 2011a). As awareness of these issues grew, so, too, did public and political outrage. Governments and intergovernmental organizations began discussing what sort of policies could be adopted at the domestic and international scales.

At the international scale, the UN took the lead. It proposed amendments to the 1989 Basel Convention (SBC, 2011b) and formally recognized WEEE as a priority issue in 2002 at Conference of the Parties (COP) (SBC, 2011c). Four years later, the COP adopted the Nairobi Declaration and "called for more structured and enhanced efforts towards achieving global solutions for management of e-waste problems" (SBC, 2011c). Alongside these developments, several regional international policy responses also emerged. The EU adopted a directive on waste electrical and electronic equipment ("WEEE Directive", Directive 2002/96/EC) and a directive restricting the use of certain hazardous substances in electrical and electronic equipment ("RoHS Directive", Directive 2002/95/EC). Twelve members from the Organisation of African Unity (now the AU) negotiated the Bamako Convention in 1991, which completely prohibited the import of hazardous and radioactive waste (SBC, 2011c). Meanwhile, in Asia, Oceania, South America, and North America, no region-wide international policies or conventions were adopted to directly address WEEE. Despite this latter lack of responses at the international scale, many countries around the world responded with diverse regulations, policies, infrastructures, and partnerships at the domestic scale (Atasu et al., 2013; Dwivedy et al., 2015; Heeks et al., 2015; Pandey & Govind, 2014; Sthiannopkao & Wong, 2013; Widmer et al., 2005).

In Canada, the responsibilities for managing and processing e-waste have been split between provincial and municipal governments via extended producer responsibility (EPR) and product stewardship programs (ECCC, 2013; Ongondo et al., 2011). These programs share WEEE costs and responsibilities between manufacturers and consumers, as well as public sector and private sector recycling facilities. Similarly, India's government recently adopted a set of extended producer responsibility rules, which took effect on 1 October 2016 and set out how manufacturers, consumers, and municipal organizations must collect and manage e-waste (Dwivedy et al., 2015; Bureau, 2016). The Chinese government has also established many rules, regulations, policies, and recycling centers across China to support WEEE processing. It also made the import of WEEE illegal (Ongondo et al., 2011). In the EU, many member states implemented their own policies, laws, and systems at the domestic scale in response to the WEEE and RoHS Directives. These policies included mandatory WEEE collection schemes for electronics retailers, WEEE recycling codes of practice, and non-profit organizations to manage WEEE recycling and take-back schemes (Ongondo et al., 2011). In the AU, many countries have also responded in diverse ways. Kenya established a public-private partnership between the government and members of the ICT community to create the East Africa Compliant Recycling facility in 2013 (Judge, 2013). South Africa passed legislation that guides the management of hazardous substances and waste, and the country is home to many formal WEEE processing facilities (E-Waste Association of South Africa, 2016; Ongondo et al., 2011). In contrast, Benin, Senegal, and Uganda are reported to lack specific

144 *Vanessa Thomas*

policies, legislation, and infrastructure to deal with WEEE (Ongondo et al., 2011; Ssekandi, 2016).

Despite these widespread and heterogeneous public policy, regulatory, and infrastructural interventions at the domestic and international scales, many global WEEE problems persist. WEEE continues to be the fastest-growing waste stream globally, and it continues to flow from the Global North to the Global South (Lepawsky, 2014; Sthiannopkao & Wong, 2013). Large, informal, and often illegal WEEE "dumping grounds" have developed in India, China, Ghana, and Nigeria, and these informal sites are damaging air quality, water sources, and human health (Pandey & Govind, 2014; SBC, 2011c; Ssekandi, 2016; Sthiannopkao & Wong, 2013). Many countries report that residents – including those who work in the dumping grounds – have little awareness of the risks associated with mishandling e-waste (Dwivedy et al., 2015; Pandey & Govind, 2014; SBC, 2011c; Ssekandi, 2016). Even when they are aware of the risks, people working in the informal dumpling grounds often have few or no other employment opportunities available to them (Pandey & Govind, 2014; Widmer et al., 2005).

WEEE policy informing HCI

WEEE policies have already directly influenced the material composition of hardware that we use in HCI. The EU's "RoHS Directive" required "heavy metals such as lead, mercury, cadmium, and hexavalent chromium and flame retardants such as polybrominated biphenyls (PBB) or polybrominated diphenyl ethers (PBDE) to be substituted by safer alternatives" (EU, 2017), and the hardware designed to meet these requirements started selling worldwide. Moreover, WEEE public policy has had a direct influence on if, how, and when people store their end-of-life electronics (Atasu et al., 2013; Dwivedy et al., 2015). Dozens of non-HCI studies have found that policy variations can "lead to an indifferent disposer who, in all likelihood, might be tempted to illegally dump [their] used products or perpetually store them" (Dwivedy et al., 2015). Several HCI studies have examined why and how people store, repair, and throw away their technologies (Odom et al., 2009; Remy & Huang, 2015); although WEEE policies were not addressed in those studies, policy variations might have unknowingly influenced the practices of participants.

HCI informing WEEE policy

There is no evidence to suggest that HCI research has informed any WEEE policies. However, many HCI studies could be of interest to policymakers looking to address the persistent failures of WEEE policies (Atasu et al., 2013; Lepawsky, 2014; Ongondo et al., 2011). In particular, studies about repair practices and cultures (e.g., Jackson & Kang (2014); Rosner et al. (2013); Sun et al. (2015)), as well as many of the studies about obsolescence (Remy & Huang, 2015), might offer policymakers new insights about the social practices affecting their WEEE policies, systems, and regulations. Moreover, some policymakers who work at the international scale favor the widespread adoption of digital technologies; they might be unaware of the health and environmental issues associated with poor WEEE management. In these cases, members of the HCI community – especially the HCI4D community – could use their interdisciplinary expertise to help inform international WEEE and sustainable development policies (e.g., Bidwell et al. (2008); Chetty & Grinter (2007); Fang et al. (2009)).

GREEN PUBLIC PROCUREMENT (GPP) POLICIES

Governments have a great deal of purchasing power, which they can use to influence the design, production, and consumption of products and services worldwide (Brammer & Walker, 2011; Günther & Scheibe, 2006; Testa et al., 2016). In some countries, governmental purchasing power fuels between 8 and 25% of the national GDP (Brammer & Walker, 2011). As concerns about environmental sustainability have grown, many levels of government have adopted "green" procurement policies (GPP) in an attempt to integrate social and environmental considerations into their purchasing processes (Brammer & Walker, 2011). These GPPs have proven to be important tools through which to stimulate eco-innovation (Beláustegui, 2011; Brammer & Walker, 2011; Günther & Scheibe, 2006; Testa et al., 2016), including in the realm of ICTs. These green ICT procurement policies often work in tandem with governmental "digital asset management" programs, which dictate how often digital technologies need to be replaced and what types of technologies should be purchased in their place (Austerberry, 2004).

Few examples of GPPs appear to exist at the international scale. The sole exception appears to be the international "Sustainable Public Procurement Programme" launched by the UN Environment Program (UNEP) in 2014 (UNEP, 2017). But many examples of GPPs exist at the domestic scale. For example, in the United States, national green ICT procurement policies were first announced in 2007 when George W. Bush issued Executive Order (EO) 13423 and required all federal agencies to use the Electronic Product Environmental Assessment Tool (EPEAT). EPEAT is a tool that offers third-party assessments of electronic devices, which "meet environmental performance criteria that address: materials selection, design for product longevity, reuse and recycling, energy conservation, end-of-life management and corporate performance" (EPA, 2016). Although EO 13423 has since been replaced, federal agencies still "prefer" environmentally sustainable electronics and EPEAT remains a popular tool within many federal departments (EPA, 2016). Dozens of other governmental organizations at the domestic scale also use EPEAT to meet their green ICT procurement needs, including the government of Canada, the Australian government, the government of the City and County of San Francisco, the government of New Zealand, Warwickshire County Council (in the UK), and the Hainan Siyuan Province's government (Green Electronics Council, 2016). However, many governments do not use EPEAT.

Some governments have designed their own unique rules, partnerships, and environmental assessment tools. For example, in Korea, the national government has a long-established "Korea Eco-label" for many products, including ICT products, and "the 'Public Procurement Minimum Green Condition Product' program went into effect to encourage green technology development" (Kim, 2014). The Malaysian government recently adopted a green IT guideline for public sector procurement (Kahlenborn et al., 2013). While there is no specific labeling system or assessment mechanism attached to the guideline, it requires "low energy consumption, minimal use of toxic material and some other environmentally related product features for ICT products procured" by the public sector (Kahlenborn et al., 2013, p. 11). In the UK, the government has a "Greening Government: ICT Strategy" that addresses the procurement, use, and disposal of computer applications, data centers, and digital devices or peripherals (HMSO, 2016).

Despite the growing prevalence of GPPs, many barriers persist with regards to their implementation and adoption at the domestic and international scales. Beláustegui (2011) identified a list of 11 barriers influencing the adoption of green procurement practices in Latin America, including a lack of information and knowledge on sustainable public procurement, insufficient offers from suppliers, and a paucity of interest in pursuing sustainable practices on the part of procurement officers. Similar barriers were found in studies from Italy (Testa et al., 2016), Korea (Kim, 2014), Germany (Günther & Scheibe, 2006), and over a dozen other countries (Brammer & Walker, 2011).

GPP policy informing HCI

GPPs, and the environmental performance requirements they helped establish, have influenced the material composition and energy consumption of technologies used and studied by the HCI community studies, especially those working with governmental organizations. Despite this influence, GPPs appear to be absent from HCI research. To date, there have been no direct studies – or even acknowledgements – of GPPs, green ICT ISOs (e.g., ISO 11469), nor EPEAT. However, many of Blevis' initial sustainable interaction design (SID) principles (Blevis, 2007) appear to align with the performance criteria set for EPEAT-certified products (e.g., they "must meet environmental performance criteria that address: materials selection, design for product longevity, reuse and recycling, energy conservation, end-of-life management and corporate performance" (Environmental Protection Agency [EPA], 2016) and there may be opportunities to develop SID further by looking at GPPs, Green ICT ISOs, and EPEAT.

HCI informing GPP policy

Many green ICT procurement policies focus primarily on hardware procurement, excluding software or digital service procurement. This oversight could be an area where the HCI community could provide expertise (Preist et al., 2016). We have access to a broad body of multidisciplinary work that examines the environmental effects of hardware, software, and digital services, and that work could be used to inform narrowly scoped green ICT procurement policies. Moreover, GPPs have only recently attracted the attention of scholars (Brammer & Walker, 2011), and few studies of green ICT procurement specifically exist. The HCI community could undertake research to help governments improve their existing green ICT procurement policies (e.g., by rethinking their asset management programs or studying how to address the paucity of interest noted by (Beláustegui, 2011)).

Engaging with EPP: opportunities and challenges

Environmental public policies are rich and complex, much like the issues they are attempting to address. Governments have adopted diverse rules, standards, guidelines, laws, and strategies to address WEEE and green procurement at the domestic and international scales, and they have faced many challenges and barriers while doing so. Similar challenges likely exist in every EPP domain that the HCI community might choose to engage with, whether that be energy efficiency, disaster and crisis response, international development, or food security policy. So how should we engage with these complex EPP domains? For what aims, and at what scales, should we engage with them? Advice about and answers to these questions come from many places, including from members of our own community.

Lazar et al. (2016) stated that "members of the HCI community need to engage, on a regular basis, with regulatory processes, at the regional, national, and multinational levels" (p. 126). For those of us interested in EPPs, this engagement could involve establishing relationships with local and national governmental departments focused on the environment, or with environmentally driven non-profit and for-profit organizations (e.g., the Green Electronics Council, the International Organization for Standardization [ISO], local WEEE recycling facilities.). We could join unions, launch environmentally focused cooperatives, consult with businesses that wish to reduce their carbon footprint (e.g., Preist et al. (2016)), write to politicians, or join protest movements. We could partner with environmental policy scholars who may already be working on relevant studies, or with scholars from other disciplines who have engaged with EPP (e.g., Shove (2010)) and tangential policy domains (e.g., Grimpe et al. (2014)). While establishing these relationships, we will likely need to consistently communicate our research findings to these communities and organisations, and incorporate their research findings into our own work.

If we choose to turn our focus inwards, then we might concentrate on highlighting the connections between EPP and HCI to the HCI community. We might choose to follow the advice of Hazas et al. (2012), who suggest we unite to collectively "pick our battles" with EPP. We could adjust our educational materials to include discussions of EPP (Lazar et al., 2016, p. 130). We could approach our professional organizations to discuss and demonstrate where and how our work intersects with environmental public policy (Lazar et al., 2016, p. 125–127). Or we could simply spend time reflecting on our own research practices and topics. For example, Silberman et al. (2014) have already suggested that we need to read beyond the HCI literature. Kraft and Furlong suggest that, to effectively engage with public policy, we should go even further by reading many non-academic sources (Kraft & Furlong, 2014, p. 3–6).

Pursuing any of these activities could be quite time-consuming (Lazar et al., 2016, p. 131). Public policymaking involves its own practices and norms, and learning to speak the right language to engage with those practices might not come easily or naturally to all of us (Freeman et al., 2011; Nathan & Friedman, 2010). As many of us know, and as Grimpe et al. (2014) remind us, "many researchers and practitioners already face various responsibilities in their daily lives and work, including numerous domain-specific requirements, and may view extra activities resulting from this [policy-focused] agenda as quite onerous". But this should not deter us. All research can be onerous; we should pursue it out of passion and an interest in shaping the world around us. The HCI community has interdisciplinary strengths that place many members in a unique position to engage with EPP at the domestic and international scales. If we put our strengths to use and seize opportunities to engage with public policy, or if we actively reach out to governmental departments, agencies, and organizations, we'll be able to build on our calls to activism and engagement.

Conclusion

Many papers within HCI have called for researchers to engage with and negotiate the complex and messy domain of environmental public policy (Goodman, 2009; Håkansson & Sengers, 2014; Knowles et al., 2014; Silberman et al., 2014) With the exception of a NordiCHI 2016 (Eriksson et al., 2016) workshop focused on the UN Sustainable Development Goals, there is little evidence to suggest that the HCI community has responded to their own calls to engage with environmental public policies. To foster discussion and inspire excitement, this chapter has explored two interconnected types of environmental public policy that might be of interest to, and benefit from engagement with, the HCI community: green procurement and WEEE public policies. Both sets of policies take many shapes and forms around the world, and their diversity – as well as the diversity of the innumerable actors who influence them – offers the HCI community many exciting opportunities for engagement. Those opportunities for engagement might not prove to be immediately easy or fruitful, in part due to the complexities of public policymaking processes; but the opportunities will persist nonetheless, and they might offer members of the HCI community a new avenue through which to enact change.

Bibliography

Atasu, A., Ozdemir, O., & Van Wassenhove, L. N. (2013). Stakeholder perspectives on e-waste take-back legislation. *Production and Operations Management, 22*(2), 382–396. Retrieved from http://dx.doi.org/10.1111/j.1937-5956.2012.01364.x

Austerberry, D. (2004). *Digital asset management*. Focal.

Bates, O., Hazas, M., Friday, A., Morley, J., & Clear, A. K. (2014). Towards an holistic view of the energy and environmental impacts of domestic media and IT. In *Proceedings of the SIGCHI Conference on Human Factors in Computing Systems* (pp. 1173–1182). New York, NY: ACM. http://doi.acm.org/10.1145/2556288.2556968

Becker, C., Chitchyan, R., Duboc, L., Easterbrook, S., Penzenstadler, B., Seyff, N., & Venters, C. C. (2015). Sustainability design and software: The Karlskrona manifesto. In *Proceedings of the 37th International Conference on Software Engineering – vol. 2* (pp. 467–476). Piscataway, NJ: IEEE Press. Retrieved from http://dl.acm.org/citation.cfm?id=2819009.2819082

Beláustegui, V. (2011). Las compras públicas sustentables en américa latina. Estado de avance y elementos clave para su desarrollo. *Organisation of American States*. Retrieved from www.oas.org/es/sap/dgpe/pub/compras2.pdf

Blevis, E. (2007). Sustainable interaction design: Invention & disposal, renewal & reuse. In *Proceedings of the SIGCHI Conference on Human Factors in Computing Systems* (pp. 503–512). New York, NY: ACM. http://doi.acm.org/10.1145/1240624.1240705

Brammer, S., & Walker, H. (2011). Sustainable procurement in the public sector: An international comparative study. *International Journal of Operations & Production Management*, *31*(4), 452–476. Retrieved from http://dx.doi.org/10.1108/01443571111119551

Bureau, E. (2016). Environment ministry notifies strict new e-waste management rules, 2016. *The Economic Times of India*. Retrieved from http://economictimes.indiatimes.com/articleshow/51528947.cms

Chetty, M., & Grinter, R. E. (2007). HCI4D: HCI challenges in the global south. In *CHI '07 Extended Abstracts on Human Factors in Computing Systems* (pp. 2327–2332). New York, NY: ACM. http://doi.acm.org/10.1145/1240866.1241002

Crivellaro, C., Taylor, A., Vlachokyriakos, V., Comber, R., Nissen, B., & Wright, P. (2016). Re-making places: HCI, 'community building' and change. In *Proceedings of the 2016 SIGCHI Conference on Human Factors in Computing Systems* (pp. 2958–2969). New York, NY: ACM. http://doi.acm.org/10.1145/2858036.2858332

Danielsen, C., Taylor, A., & Majerus, W. (2011, January). Design and public policy considerations for accessible e-book readers. *interactions*, *18*(1), 67–70. Retrieved from http://doi.acm.org/10.1145/1897239.1897254

Davis, J., Hochheiser, H., Hourcade, J. P., Johnson, J., Nathan, L., & Tsai, J. (2012). Occupy CHI!: Engaging U.S. policymakers. In *CHI '12 Extended Abstracts on Human Factors in Computing Systems* (pp. 1139–1142). New York, NY: ACM. http://doi.acm.org/10.1145/2212776.2212406

DiSalvo, C., Sengers, P., & Brynjarsdóttir, H. (2010). Mapping the landscape of sustainable HCI. In *Proceedings of the SIGCHI Conference on Human Factors in Computing Systems* (pp. 1975–1984). New York, NY: ACM. http://doi.acm.org/10.1145/1753326.1753625

Dwivedy, M., Suchde, P., & Mittal, R. (2015). Modeling and assessment of e-waste takeback strategies in India. *Resources, Conservation and Recycling*, *96*, 11–18. Retrieved from www.sciencedirect.com/science/article/pii/S092134491500004X

Dye, T. (1978). *Understanding public policy* (3rd ed.). Don Mills, ON, Canada: Pearson.

Elsevier, B. V. (2017). *Environmental science and policy*. Retrieved from www.journals.elsevier.com/environmental-science-and-policy/

Environment and Climate Change Canada. (2013). *Extended producer responsibility*. Retrieved from www.ec.gc.ca/gdd-mw/default.asp?lang=Enn=FB8E9973-1

Environmental Protection Agency. (2016). *Electronic Product Environmental Assessment Tool (EPEAT)*. Retrieved from www.epa.gov/greenerproducts/electronic-product-environmental-assessment-tool-epeat

Eriksson, E., Pargman, D., Bates, O., Normark, M., Gulliksen, J., Anneroth, M., & Berndtsson, J.(2016). HCI and UN's sustainable development goals: Responsibilities, barriers and opportunities. In *Proceedings of the 9th Nordic conference on human-computer interaction* (pp. 140:1–140:2). New York: ACM. Retrieved from http://doi.acm.org/10.1145/2971485.2987679 doi:10.1145/2971485.2987679

European Union. (2017). *The ROHS directive*. Retrieved from http://ec.europa.eu/environment/waste/rohseee/index en.htm

E-Waste Association of South Africa. (2016). *Home*. Retrieved from http://http://ewasa.org/
Fang, A. T., Khozein, R., Mendez-Baiges, S. M., & Shin, E. Y. (2009). eXtend: Reducing e-waste through redistribution of local IT resources. In *CHI '09 Extended Abstracts on Human Factors in Computing Systems* (pp. 2817–2822). New York, NY: ACM. http://doi.acm.org/10.1145/1520340.1520408
Fantauzzacoffin, J., Berzowska, J., Edmonds, E., Goldberg, K., Harrell, D. F., & Smith, B. (2012). The arts, HCI, and innovation policy discourse: Invited panel. In *CHI '12 Extended Abstracts on Human Factors in Computing Systems* (pp. 1111–1114). New York, NY: ACM. http://doi.acm.org/10.1145/2212776.2212399
Freeman, R., Griggs, S., & Boaz, A. (2011). The practice of policy making. *Evidence & Policy: A Journal of Research, Debate and Practice, 7*(2), 127–136. Retrieved from www.ingentaconnect.com/content/tpp/ep/2011/00000007/00000002/art00001
Froehlich, J., Findlater, L., & Landay, J. (2010). The design of eco-feedback technology. In *Proceedings of the SIGCHI Conference on Human Factors in Computing Systems* (pp. 1999–2008). New York, NY: ACM. http://doi.acm.org/10.1145/1753326.1753629
Günther, E., & Scheibe, L. (2006). The hurdle analysis: A self-evaluation tool for municipalities to identify, analyse and overcome hurdles to green procurement. *Corporate Social Responsibility and Environmental Management, 13*(2), 61–77. Retrieved from http://dx.doi.org/10.1002/csr.92 doi: 10.1002/csr.92
Goodman, E. (2009). Three environmental discourses in human-computer interaction. In *CHI '09 Extended Abstracts on Human Factors in Computing Systems* (pp. 2535–2544). New York, NY: ACM. http://doi.acm.org/10.1145/1520340.1520358
Green Electronics Council. (2016). *EPEAT purchasers*. Retrieved from www.epeat.net/participants/purchasers/
Grimpe, B., Hartswood, M., & Jirotka, M. (2014). Towards a closer dialogue between policy and practice: Responsible design in HCI. In *Proceedings of the SIGCHI Conference on Human Factors in Computing Systems* (pp. 2965–2974). New York, NY: ACM. http://doi.acm.org/10.1145/2556288.2557364
Gulliksen, J., von Axelson, H., Persson, H., & Göransson, B. (2010, May). Accessibility and public policy in Sweden. *Interactions, 17*(3), 26–29. Retrieved from http://doi.acm.org/10.1145/1744161.1744168
Håkansson, M., & Sengers, P. (2014). No easy compromise: Sustainability and the dilemmas and dynamics of change. In *Proceedings of the 2014 Conference on Designing Interactive Systems* (pp. 1025–1034). New York, NY: ACM. http://doi.acm.org/10.1145/2598510.2598569
Hazas, M., Brush, A. J. B., & Scott, J. (2012, September). Sustainability does not begin with the individual. *Interactions, 19*(5), 14–17. Retrieved from http://doi.acm.org/10.1145/2334184.2334189
Hazas, M., Morley, J., Bates, O., & Friday, A. (2016). Are there limits to growth in data traffic? On time use, data generation and speed. In *Proceedings of the second workshop on computing within limits* (pp. 14:1–14:5). New York, NY: ACM. Retrieved from http://doi.acm.org/10.1145/2926676.2926690
Heeks, R., Subramanian, L., & Jones, C. (2015). Understanding e-waste management in developing countries: Strategies, determinants, and policy implications in the Indian ICT sector. *Information Technology for Development, 21*(4), 653–667. Retrieved from http://dx.doi.org/10.1080/02681102.2014.886547
Her Majesty's Stationary Office. (2012). *Sustainable procurement: The GBS for office ICT equipment*. Retrieved from www.gov.uk/government/publications/sustainable-procurement-the-gbs-for-office-ict-equipment

Her Majesty's Stationary Office. (2016). *Government Information Technology (IT) – Greening Government ICT strategy*. Retrieved from www.gov.uk/government/publications/greening-government-ict-2015-annual-report

Her Majesty's Stationary Office. (2017). *Policies*. Retrieved from www.gov.uk/government/policies

Howlett, M., & Ramesh, M. (2003). *Studying public policy: Policy cycles and policy subsystems* (2nd ed.). Don Mills, ON, Canada: Oxford University Press.

Jackson, S. J., & Kang, L. (2014). Breakdown, obsolescence and reuse: HCI and the art of repair. In *Proceedings of the SIGCHI Conference on Human Factors in Computing Systems* (pp. 449–458). New York, NY: ACM. http://doi.acm.org/10.1145/2556288.2557332

Judge, P. (2013). Giant Kenyan e-waste scheme hopes to clean up toxic tech. *Tech Week Europe*. Retrieved from www.techweekeurope.co.uk/workspace/e-waste-kenya-dump-green-134350

Kahlenborn, W., Mansor, N., & Adham, K. N. (2013). Government green procurement (GGP) short-term action plan 2013–2014. *Sustainable Consumption and Production Malaysia*. Retrieved from www.scpmalaysia.gov.my/images/GGP%20short%20term%20action%20Plan%20-%20250613%20-%20final0.pdf

Kim, D.-I. (2014). Korea. In V. Lember, R. Kattel, & T. Kalvet (Eds.), *Public procurement, innovation and policy: International perspectives* (pp. 191–211). Berlin, Heidelberg: Springer. http://dx.doi.org/10.1007/978-3-642-40258-610

Knowles, B., Blair, L., Coulton, P., & Lochrie, M. (2014). Rethinking Plan A for sustainable HCI. In *Proceedings of the SIGCHI Conference on Human Factors in Computing Systems* (pp. 3593–3596). New York, NY: ACM. http://doi.acm.org/10.1145/2556288.2557311

Kraft, M., & Furlong, S. (2014). *Public policy: Politics, analysis, and alternatives*. Sage.

Lazar, J., Abascal, J., Barbosa, S., Barksdale, J., Friedman, B., Grossklags, J., . . . Wentz, B. (2016). Human-computer interaction and international public policymaking: A framework for understanding and taking future actions. *Foundations and Trends® Human-Computer Interaction*, *9*(2), 69–149. Retrieved from http://dx.doi.org/10.1561/1100000062

Lepawsky, J. (2014). The changing geography of global trade in electronic discards: Time to rethink the e-waste problem. *The Geographical Journal*. Retrieved from http://dx.doi.org/10.1111/geoj.12077

Lund, A., Scholtz, J., & Bevan, N. (2012, January). Why the CHI community should be involved in standards: Stories from three CHI participants. *Interactions*, *19*(1), 70–74. Retrieved from http://doi.acm.org/10.1145/2065327.2065341

Mirnig, A. G., Meschtscherjakov, A., Wurhofer, D., Meneweger, T., & Tscheligi, M. (2015). A formal analysis of the ISO 9241–210 definition of user experience. In *Proceedings of the 33rd Annual ACM Conference Extended Abstracts on Human Factors in Computing Systems* (pp. 437–450). New York, NY: ACM. http://doi.acm.org/10.1145/2702613.2732511

Nathan, L. P., & Friedman, B. (2010, September). Interacting with policy in a political world: Reflections from the voices from the Rwanda tribunal project. *Interactions*, *17*(5), 56–59. Retrieved from http://doi.acm.org/10.1145/1836216.1836231. doi:10.1145/1836216.1836231

Odom, W., Pierce, J., Stolterman, E., & Blevis, E. (2009). Understanding why we preserve some things and discard others in the context of interaction design. In *Proceedings of the SIGCHI conference on human factors in computing systems* (pp. 1053–1062). New York: ACM. Retrieved from http://doi.acm.org/10.1145/1518701.1518862 doi:10.1145/1518701.1518862

Ongondo, F., Williams, I., & Cherrett, T. (2011). How are WEEE doing? A global review of the management of electrical and electronic wastes. *Waste Management*, *31*(4), 714–730. Retrieved from www.sciencedirect.com/science/article/pii/S0956053X10005659

Pandey, P., & Govind, M. (2014). Social repercussions of e-waste management in India: A study of three informal recycling sites in Delhi. *International Journal of Environmental Studies*, *71*(3), 241–260. Retrieved from www.tandfonline.com/doi/abs/10.1080/00207233.2014.926160

Penzenstadler, B., Bauer, V., Calero, C., & Franch, X. (2012, May). Sustainability in software engineering: A systematic literature review. In *16th International Conference on Evaluation Assessment in Software Engineering* (ease 2012) (pp. 32–41). doi: 10.1049/ic.2012.0004

Pierce, J., & Paulos, E. (2012). Beyond energy monitors: Interaction, energy, and emerging energy systems. In *Proceedings of the SIGCHI Conference on Human Factors in Computing Systems* (pp. 665–674). New York, NY: ACM. http://doi.acm.org/10.1145/2207676.2207771

Preist, C., Schien, D., & Blevis, E. (2016). Understanding and mitigating the effects of device and cloud service design decisions on the environmental footprint of digital infrastructure. In *Proceedings of the SIGCHI Conference on Human Factors in Computing Systems* (pp. 1324–1337). New York, NY: ACM. http://doi.acm.org/10.1145/2858036.2858378

Remy, C., & Huang, E. M. (2015). Addressing the obsolescence of end-user devices: Approaches from the field of sustainable HCI. In L. M. Hilty & B. Aebischer (Eds.), *ICT innovations for sustainability* (Vol. 310, pp. 257–267). Springer. http://dx.doi.org/10.1007/978-3-319-09228-7_15

Rosner, D. K., Jackson, S. J., Hertz, G., Houston, L., & Rangaswamy, N. (2013). Reclaiming repair: Maintenance and mending as methods for design. In *CHI '13 Extended Abstracts on Human Factors in Computing Systems* (pp. 3311–3314). New York, NY: ACM. http://doi.acm.org/10.1145/2468356.2479674

Secretariat of the Basel Convention. (2011a). *Convention overview*. Retrieved from http://basel.int/TheConvention/Overview/tabid/1271/Default.aspx

Secretariat of the Basel Convention. (2011b). *E-waste overview*. Retrieved from http://basel.int/Implementation/Ewaste/Overview/tabid/4063/Default.aspx

Secretariat of the Basel Convention. (2011c). *Where are WEEE in Africa? Findings from the Basel Convention E-waste Africa Programme*. Retrieved from www.basel.int/Portals/4/download.aspx?d=UNEP-CHW-EWASTE-PUB-WeeAfricaReport.English.pdf

Shove, E. (2010). Beyond the ABC: Climate change policy and theories of social change. *Environment and Planning A*, *42*(6), 1273–1285. Retrieved from http://epn.sagepub.com/content/42/6/1273.abstract

Silberman, M. S., Nathan, L., Knowles, B., Bendor, R., Clear, A., Håkansson, M., . . . Mankoff, J. (2014, September). Next steps for sustainable HCI. *Interactions*, *21*(5), 66–69. Retrieved from http://doi.acm.org/10.1145/2651820

Ssekandi, J. (2016). Uganda: Environmentalists alarmed over growing e-waste. *The Observer (Kampala)*. Retrieved from http://allafrica.com/stories/201609020699.html

Sthiannopkao, S., & Wong, M. H. (2013). Handling e-waste in developed and developing countries: Initiatives, practices, and consequences. *Science of the Total Environment*, *463464*, 1147–1153. Retrieved from www.sciencedirect.com/science/article/pii/S0048969712009217

Sun, Y., Lindtner, S., Ding, X., Lu, T., & Gu, N. (2015). Reliving the past & making a harmonious society today: A study of elderly electronic hackers in China. In *Proceedings* of the 18th ACM Conference on Computer Supported Cooperative Work & Social Computing (pp. 44–55). New York, NY: ACM. http://doi.acm.org/10.1145/2675133.2675195

Testa, F., Grappio, P., Gusmerotti, N. M., Iraldo, F., & Frey, M. (2016). Examining green public procurement using content analysis: Existing difficulties for procurers and useful recommendations. *Environment, Development and Sustainability*, *18*(1), 197–219. http://dx.doi.org/10.1007/s10668-015-9634-1

Thomas, V., Remy, C., Hazas, M., & Bates, O. (2017). HCI and environmental public policy: Opportunities for engagement. In *Proceedings of the 2017 CHI Conference on Human Factors in Computing Systems* (pp. 6986–6992). New York: ACM.

United Nations Environment Programme. (2017). *Sustainable public procurement*. Retrieved from http://web.unep.org/10yfp/programmes/sustainable-public-procurement

United Nations University. (2013). *UNU releases China e-waste study*. Retrieved from http://unu.edu/news/news/unu-releases-china-e-waste-study.html

Vlachokyriakos, V., Crivellaro, C., Le Dantec, C. A., Gordon, E., Wright, P., & Olivier, P. (2016). Digital civics: Citizen empowerment with and through technology. In *Proceedings of the 2016 CHI Conference Extended Abstracts on Human Factors in Computing Systems* (pp. 1096–1099). New York, NY: ACM. http://doi.acm.org/10.1145/2851581.2886436

Widmer, R., Oswald-Krapf, H., Sinha-Khetriwal, D., Schnellmann, M., & Bni, H. (2005). Global perspectives on e-waste. *Environmental Impact Assessment Review*, *25*(5), 436–458. Retrieved from www.sciencedirect.com/science/article/pii/S0195925505000466

9 On the inherent contradictions of teaching sustainability at a technical university

Elina Eriksson and Daniel Pargman

On the necessity of rocking the boat

Computers and digitalization have greatly shaped our world and are now an unavoidable part of modern society. Weiser's (1991) vision of ubiquitous computing has in many respects not only been met but has in affluent parts of the world been surpassed (Bell & Dourish, 2007). Digital artifacts and devices surround us and have invisibly and seamlessly permeated everything we do. Our modern societies are however not sustainable. We have overstepped several planetary boundaries and risk overstepping several more (Steffen et al., 2015). We are about to reach limits as to the resources we can extract from the earth (Bardi, 2014), and the changes wreaked are by now so profound that they will likely last for a geological period of time (Steffen et al., 2007). In light of this, it is of utmost importance to strive towards a sustainable society, and this is a responsibility that falls on many disciplines and sectors. We believe that engineering students could be key drivers in this change since many will eventually enter positions of power from which they will make decisions that will shape our future society.

At KTH Royal Institute of Technology, Sweden, we have, during the last five years, been responsible for teaching an introductory course in sustainability to our media technology engineering students (Eriksson & Pargman, 2014). Even though ICT in itself has a substantial environmental impact (Cramer, 2012), it supposedly also harbors a large potential to abate environmental impacts in *other* sectors (Neves et al., 2012). Proposed solutions to present and future sustainability challenges must be commensurable with the magnitude of the challenges we face, and this is naturally exceedingly difficult to cover in a single university course (being part of a five-year-long program). The easier and more palatable alternative is to adopt more modest goals and focus on smaller solutions, a perspective we have earlier termed "vanilla sustainability" (Pargman & Eriksson, 2013). This perspective is captured in the ethos of *"every little bit helps"* but has been severely criticized by, for example, MacKay, who counters by stating that *"if everyone does a little, we'll achieve only a little"* (McKay, 2009, p. 3).

We are convinced that we instead need to teach our students a more radical perspective on sustainability, a perspective we have previously referred to as "strong sustainability" (Pargman & Eriksson, 2016). However, taking this stance when teaching sustainability does come at a cost, and we as teachers must consequently be ready to be challenged, together with the structures around us.

In this book chapter, we present five anecdotes that exemplify occasions when our course has run up to various barriers and has come into conflict with everyday beliefs, habits, and expectations of students, colleagues, and others. The anecdotes will unveil tensions and contradictions of various kinds, *but these might constitute an unavoidable precondition for change*. We believe that the need to change society naturally starts at the university and in our classroom. There just might not be an easy and straightforward answer for how to handle (or "fix") the tensions and contradictions we meet. We rather suggest that we must learn to live with them and to work through them. In this sense they resemble *predicaments* that we have to *accommodate* rather than *problems* that once and for all can be *solved* (Greer, 2008). Admittedly we do not know the correct answer for how to handle these challenges. However, we hope that by describing and discussing examples of challenges we have faced, others will realize they are not the first to encounter incomprehension, push-back, and negative reactions. That can in itself offer some degree of solace and we suggest the next step is to compare notes and single out successful strategies to overcome these and other challenges.

What happens when you rock the boat

In this section, we will give examples of different types of friction that have occurred when teaching our students a strong sustainability perspective. We will do this by presenting five anecdotes, with each followed by an elaboration. Each anecdote is described from a first-person perspective as experienced by one or the other of the two authors of this chapter.

On loss, fear and sadness

> It's a sunny Friday afternoon, but the lecture hall is dark and cool. I am wrapping up my lecture, describing what Earth might look like in a worst-case scenario – if the average temperature on the planet was six degrees warmer as a result of climate change. The slide-deck ends with a black slide and I smile a slightly ironic smile and wish the audience a great weekend. I feel rather shaken but relieved. After weeks of reading up on planetary boundaries, climate change, ice core data, sea level rise, and species extinction, I am now finished. When I pack up my computer, I see three students approaching the lectern in the corner of my eye. As I turn to them, I register the crossed arms, as if they are grasping for support, and that the student in the middle is teary-eyed. One of them asks without prompting: "Can't you say something more optimistic?"

As a lecturer, I stood dumbfounded. In this lecture, I had presented up-to-date facts: all the measurements, calculations, all the observable changes of the planetary system. I guess it should not have been surprising that some students' reactions to these facts were not only intellectual but also deeply emotional. I had myself, over the weeks that I prepared for the lecture, felt an increasing sense of alarm, dread, and sorrow over the state of the world and the insurmountable predicaments we are facing. Not to mention the anger and frustration I felt for the lack of concern from politicians, industry, and media. Instead of admitting my own apprehension, I tried to say something chirpy along the lines of *"we'll talk about what possible solutions might look like later in the course"* and quickly bid the students goodbye and scurried away. I carried a huge emotional backpack when I joined my colleagues for a coffee break later in the afternoon, and as I described the incident to them. Had I done something wrong, evoking such feelings in my students? One senior colleague answered that yes, I should not dump something like that onto them and especially not on a Friday afternoon.

But what was I really "dumping" something on them? I had but accounted for a score of scientific facts (Stocker et al., 2013; Steffen et al., 2015; Steffen et al., 2007; Füssel & Jol, 2012), facts that we in every other situation revere at a technical university. But in light of this anecdote, further questions arise. Should we avoid evoking emotions – both my own and those of my students? Should I instead have wrapped up my students in cotton wool and downplayed the scale and the urgency of the problems we are facing? If so, how exactly am I supposed to do that? By portioning out (moderately) bad news in between cheerful accounts of what we are currently working on, might that help? If it, on the other hand, is fine – or even commendable – to evoke emotions in my students, exactly what responsibilities do I then have as their teacher? Should I take care of their emotions, and how exactly am I supposed to do that? For heaven's sake, I am a university teacher and not a therapist, and our seminars are *academic* seminars and not support groups! Again, what am I supposed to do? Direct them to the nearest health center or tell them to talk to a psychologist about their climate angst? Deliver a bunch of facts in a detached manner and let them work it out and deal with it as grown-ups as best as they can? Or should we embrace their worries and follow up the lectures with some structure that makes it possible for the students to vent their concerns?

As shown in this section, emotions are stirred when presenting facts about our current situation, which on the other hand might be necessary in order to instigate action in our students (Weber, 2010). However, there are many emotional barriers that arise when approaching facts about planetary boundaries. Norgaard (2011) shows that the most common emotion management strategies for avoiding fear and helplessness are to control the exposure to information, to not think far ahead, and to focus on the things that one *can* do (but that perhaps are not the most effective). Unfortunately, educators were one of the groups that most frequently used these strategies (Norgaard, 2011). Furthermore, this is not only a problem in education, but the same fear-reducing strategies are also present within the research community (Knowles & Eriksson, 2015) and when they are put aside, researchers run the risk of becoming labeled "alarmists" who can easily be ignored (Greer, 2013). But there *is* no more time and we *have* to make climate change and other planetary threats a front-of-the-mind issue (Giddens, 2009). If we ignore all negative feelings, we instead run the risk of exhibiting behavior that could potentially worsen the situation such as denial, manic activity (increased consumption), idealization of past times, and engagement in false solutions (Randall, 2009). We have to find strategies to handle negative emotions, and to talk about the hard facts, even if it hurts.

On being a difficult colleague

> My colleague stops me in the corridor with an apologetic smile pasted on his face. I am not exactly in a hurry, but I can't really talk for very long. He starts with a few pleasantries and then turns to the issue at hand: "First of all, I have to say that it is really good that you teach that sustainability course, it is sorely needed, but. . . ." I then get a tale of students who question the contents of his interaction design course, and more specifically the project suggestions that the students have been provided with. "It's not that there is anything wrong with focusing on sustainability," he tells me. It's just that none of the project ideas that he had provided seem to fit the students, who were at a loss. Maybe we could talk about this some other time? My first reaction is to answer "Good, good, that's actually great and exactly what we are working towards in our course," but I quickly pull on my concerned face and agree that this is something we have to think about and talk more about at some later point in time.

As suggested above, we were at first *happy* that our students had started to question the content in other courses. One of the goals for our course has been to engage the students and to make such an impact that they will carry the (strong) sustainability perspective with them into other courses. However, why should they need to question content in other courses in the first place? Sustainability is a prioritized perspective at our university, and since 2011 all Master of Science in engineering educational programs should, at a minimum, include at least one sustainability course (Pargman & Eriksson, 2013). The overall long-term objective is to integrate the subject in all levels of the education. Integrating sustainability in higher education can take different forms, and our case mostly resembles a centralized approach with one or a couple of dedicated courses in sustainability in the curricula (Mann et al., 2010).

Another more fundamental and radical approach is to transform the whole educational system and elevate the goal of a sustainable society to be the main aim of the education (Sterling, 2004). However, sustainability is not yet fully integrated into our educational program, and as the anecdote clearly points at, our students are thus caught in the crossfire between having one dedicated course about sustainability and having the rest of the courses and their contents remaining the way they have always been. There is a contradiction between the university's high-flying goals of sustainability and the practical goal of integrating such a perspective in the actual education. Our students can be ambassadors for change, challenging the barriers that stand in the way. One such barrier is the teachers who are responsible for other courses. In this case, the teacher in question was relatively accommodating, being fairly environmentally aware, but in need of help to formulate relevant project ideas for the students. Other teachers might be uninterested in meeting the students' concerns or even repudiating the need to include sustainability in their course contents. Another barrier is the already chock-full curriculum where few educators feel that they have room for more course content, or are ready to drop content from their courses in order to include sustainability aspects. One way to squeeze in sustainability "sideways" is to use examples that have a bearing on sustainability, such as looking at efficient algorithms and how these can save energy in a programming course. However, without a deeper grounding in sustainability discourse, such examples might be perceived as superficial and run the risk of discouraging students who are invested in the subject.

On rethinking your future career

> A student who has not particularly distinguished himself in class wants to talk to me as the course is winding down. He is a big guy and towers over me in the corridor. He has something of a slow and terse way of speaking, so I am guessing he comes from somewhere up north. I remember him from the seminars; he didn't say a lot, but whenever he said something it was interesting and to the point. He seems to be quite a thoughtful student, but not one that I feel I have come to know through the course. I am amazed when he says, "Before I took this course, I imagined I would work with something having to do with media and marketing, but now I feel I want to do something important instead." He is afraid he has chosen the wrong career. He asks me for advice about what to do. I then-and-there come to realize I don't really think I have any advice that I can give him.

Changing your worldview happens seldom, for some perhaps never. However, we believe we reach precisely that goal with one or a few students every year (although this is difficult to ascertain since students do not regularly look us up for heart-to-heart conversations about how our course has affected them). But having students adopt an alternative worldview is also where our problems start, because what do you tell a student who says that had the course been given two or three years ago, she would have switched to another educational program? Quitting our media technology engineering program, or even quitting engineering altogether has been mentioned by students of ours. So should we listen to what they say and direct them to permaculture courses, to climate activism, or to take time off their studies to become a WWOOFer (voluntary workers on organic farms)?[1] Let us imagine that we would do that, which might help the students, but would make us unpopular among the faculty and the KTH management. Also, there are financial incentives – our university is compensated based on how many students we teach and how many students pass our courses and our educational programs.

Even if we personally sympathize with their initial "panic reaction" (having gone through that phase ourselves), we have come to the conclusion that it would be grossly irresponsible of us to encourage them to make sudden career shifts – something they might regret at a later time. Having discussed this extensively, we have come to the conclusion that our official standpoint has to be that we do not just need organic farmers or full-time climate activists, but that we also need sustainable media/computer mavericks who will work towards transforming society in a sustainable direction (Mann, 2011). This perspective resonates with the writing of Fry (2007) on redirective practices, where the author argues that we need to disclose *"how practice is being determined and then uncoupling, modifying, remaking and reframing it"* (Fry, 2007, p. 9). Hence, instead of becoming a full-time climate activist, our advice is that our students should "become climate activists" by *supporting* climate activists through by developing robust low-cost systems for coordination and communication (Shirky, 2008) as well as developing their social media policy (Terranova & Donovan, 2013) and supporting their long-term digital support structures (Silberman, 2015).

On not meeting students' expectations

> I meet the guest lecturer outside the entrance and show him to the lecture hall. I am excited at having snagged someone from this particular multinational tech company to come and present their sustainability research to our students. Most students have shown up despite this not being a mandatory session and I feel proud of them. The computer is set up and we are off. Only a few minutes into the lecture I start to cringe in my front-row seat and I hear students start to whispers behind me. When the lecturer brings up a picture of a design concept – a quantified wifi-connected cutting board that can help you find recipes based on what vegetable you just put on it – well, then I just want to sink through the floor in embarrassment. This particular example of how ICT could "help make the world more sustainable" became a standing joke at our seminars for the duration of that course.

Perhaps the quantified cutting board was the lecturer's way of showing off some cool technology to impress our students and gain credibility. The lecturer might not even have particularly *liked* the quantified cutting board, but chose to present it in the belief that the students would. It just did not land very well in a group of students who had just waded through a pile of discouraging facts about climate change, peak oil, and planetary boundaries. The guest lecturer also presented other examples that were more interesting and more suitable to the scale of the challenges that sustainability presents us with, although the general gist of the lecture was not really what we had expected. The students later expressed their disappointment and their disillusionment in industry's willingness and ability to "do their bit" in the transition to a sustainable society. We had no words in defense.

Perhaps we had a stroke of bad luck, and someone more qualified could have done the company better justice in relation to their sustainability work. However, there is also criticism within the field of HCI that we are not doing enough, or that we are not doing the right things (Brynjarsdottir et al., 2012; Silberman et. al., 2014). To the man with a hammer, everything looks like a nail. In an artifact-centered discipline, which human-computer interaction is, the solution to any problem is a proposal for a new artifact. It might not be the best or the most appropriate solution, but it is the solution an HCI person will most often suggest (Baumer & Silberman, 2011). However, there is growing interest in industry for computing students who understand and want to work with sustainability-related issues. Our students could become "tempered radicals", working within the establishment, but with an activist agenda of changing the system from within (Meyerson & Scully, 1995). For this to happen, their interest in sustainability must be maintained, while they at the same time are able to work within the system and push for change. This is a position that can be difficult to maintain. Furthermore, our students will, for the most part, begin at entry-level positions in various organizations, and will initially have little influence over higher-level decision processes. If they make too large a ruckus, they will undermine their own careers and never reach the decision-making levels, so this path is a balancing act.

On facts as ideology

> This year we had the benefit of having a supremely knowledgeable student in the classroom. This particular student had worked politically in a right-of-center political party. The student fervently believed that a sustainable world could only be built on a foundation of liberalism, capitalism, free trade and economic growth. This perspective clashed with some of the course readings and lecture contents. This made for interesting seminar discussions throughout the course, for example when the student characterized the Fairtrade certification scheme as deplorable since it benefited smallholders and discriminated against larger, more economically effective producers in developing countries. During the final lecture in the course, the very same student complained that several of the course texts had been too "ideological".

We as educators first became quite defensive, feeling that we, of course, are not promoting any particular ideology. We are researchers and teachers, wardens of facts and The Truth. But, with some distance to the accusation, we can see that there are issues here worth discussing. We believe that one purpose of our course is to make the students reflect on, and question, the state of the world, since it obviously is not sustainable. Parts of the course contents consist of objective facts that are in line with what is scientifically known today. Parts per million of CO_2 in the atmosphere and acidification and eutrophication of the oceans are measurable facts. But beyond such measurements, sustainability does indeed become a value-laden subject where different political ideologies unavoidably clash (Dobson, 2007). We have reflected on this earlier, and have taken certain measures to handle the tension that exists between facts and values (Pargman & Eriksson, 2013). For example, we decided to be as open as possible with our own values and early in the course present the reasons why we personally find this topic interesting and important. Furthermore, we have several seminars where students formulate and propose questions to discuss (based on the course literature). We also strive for having a large number of guest lectures, including representatives from industry, to showcase different perspectives on sustainability.

On the other hand, is not all education to some degree an exercise in ideology creation, or at least value-laden, even though we try to make it as objective as possible? All education aims at shaping professionals that can build and maintain our society. Hence, most of what we do is colored by what we think constitutes a good society. Democracy, human rights, ethics, and now sustainability.

The hyperloop in the room

All five anecdotes above chisel out tensions and situations where we as educators have felt at a loss. At first glance, the anecdotes can seem disparate, but they all in one way or another suggest that we, as university teachers, are not sufficiently prepared to unveil and discuss values and norms, or for handling reactions from students that lie outside the domain of learning "objective" facts. As university teachers, we are not used to handling students' emotions such as fear, loss, and grief. The concrete topic of each anecdote varies; however, it is possible to argue that all the anecdotes are connected, that they represent inherent contradictions and that these anecdotes all allude to an elephant in the room (Zerubavel, 2006).

Some will be quick to suggest that the solutions to our current sustainability problems are to be found in new revolutionary technologies, which brings hope to our engineering students – be it geoengineering, breeder or fusion reactors, or of building a hyperloop that would make a short shrift of the time needed to travel between San Francisco and Los Angeles (SpaceX, 2013). And perhaps best of all, technical universities can be at the forefront of making this wonderful new high-tech world come true!

But however revolutionary the idea of the hyperloop is, we note that not one but two versions of the hyperloop were proposed in the original white paper (SpaceX, 2013). The passenger-only 6-billion-dollar version was also complemented with the larger 7.5-billion-dollar version for transporting passengers *and* their vehicles between San Francisco and Los Angeles. So even in this, the most advanced of futures, we can still not let go of the legacy of the car society (Dennis & Urry, 2009) despite the fact that polluting cars are an important driver behind our current woes! Let the contradiction sink in. The hyperloop has been framed as a disruptive innovation that would revolutionize traveling between various cities (routes have since been proposed between Toronto and Montreal, Stockholm and Helsinki, Paris and Amsterdam, etc.). It has also been touted as a supremely energy-efficient way of traveling. But to imagine that people from San Francisco would leave their cars at home and rent a car in Los Angeles (or make do without a car) is apparently too weird to propose, despite the fact that a car can easily weigh 20 times more than the person driving the car (e.g., 1,500 kg vs. 75 kg). By extension, we can assume that the idea that we could ever build a future high-tech hyperloop society *without cars* is apparently a much more outlandish proposition than the idea of the hyperloop itself!

We see here outcomes of a narrow focus on solving concrete problems, but without first asking if the right problem is being solved. Management guru Peter Drucker is well known for having said that *efficiency* is doing things right, but that *effectiveness* is doing the right things. Drucker is also known for having said that *"there is nothing quite so useless as doing with great efficiency something that should not be done at all"*. We can discern yet another tension here because engineering is about solving problems and the highest praise we can give our engineering students is that they are "problem-solvers". So what problems do engineers solve? The answer is that for engineers to be able to solve a problem, that problem first has to be defined and delimited and that there is an implicit focus on a *technological* solution. However, there is a risk that this will delimit a problem in such a way that one arrives at an efficient – but not necessarily an effective – solution (Pargman et al., 2016). That conundrum is intrinsically hard when it comes to sustainability-related issues. Problems having to do with global climate change (and with modifying natural systems) have been described not only as "wicked problems" (Rittel & Webber, 1973) but as "super wicked problems" (Levin et. al., 2012), creating situations where *"traditional analytical techniques are ill equipped to identify solutions, even when it is well recognized that actions must take place soon to avoid catastrophic future impacts"* (Levin et al., 2012, p. 123).

The belief (or the hope) that technological solutions can solve social ills is widespread and probably more so at a technical university and among engineers than among the general population. The hope for an unproblematic techno-fix is alluring, despite a mixed track record (to say the least). However, the effects of technologies are hard to trace over space and time, and the effects are furthermore often *both* positive and negative (Huesemann & Huesemann, 2011). What it comes down to in the end is a matter of beliefs and a clash between two grand narratives and two different worldviews. Many people, including many engineering students, take for granted that the future can be extrapolated from the present and the recent past. Furthermore, this imagined future is based on a narrative of expanding borders and scientific progress, growing economic prosperity and human well-being. The notion of scientific and economic progress in small incremental steps forever, or at least for a very long time, is part of the fabric of the grandest of our modern narratives and an element of the very operating system of modern societies, not to say our technological universities.

A growing number of scientists (Jackson, 2009), activists (Monbiot, 2007), and scientist-activists (Lovelock, 2010), on the other hand, point at the triple crisis (ecology, economy, energy) or aspects of it, and take as their starting points a range of different biophysical and economic limitations. The alternative narrative is one of human hubris, of running up to various limitations (energy, water, food, fish in the sea, a stable climate, etc.) and perhaps also of decline (Tainter, 1990). The anecdotes above can be seen as examples of a collision between two different worldviews – a collision between "the cornucopian paradigm" (Preist et al., 2016) and the idea that there are limits to growth (Meadows et al., 1972). When our anecdotes are pressed into service in a battle between worldviews, it is easy to see that there are no simple solutions that would help us avoid future conflicts (and the creation of more anecdotes).

Reverberations

It does not matter which specific topic we start discussing at seminars with our students because the process follows a repetitive pattern. We start to discuss sustainability, production, transportation, greenhouse gas emissions, ICT services, energy use, overpopulation, inequality, planned obsolescence, or our current economic system and then quickly have to untangle multiple interconnected strands of a larger system of intertwined issues. As the dominant narrative ("all is fine and things are getting better") becomes besieged and starts to unravel, we and our students are plunged into a rabbit hole where we suddenly see the ordinary world through a series of perspective-changing funny mirrors. This recurring exercise can unleash various emotions in our students including grief, anger, and at times hope. We have earlier discussed if what we are doing really is appropriate since the strong sustainability perspective we present *"runs the risk of shocking, depressing or even paralyzing students. . ., but can however paradoxically also be liberating, since it takes students' worries seriously instead of glossing over fundamental problems"* (Pargman & Eriksson, 2013, p. 501).

We as teachers can feel dispirited by seeing our students' reactions, by having to take on the role of being the messengers of bad news, by not having clear and simple answers about what to do, and thus by failing to guide our students through a quagmire of bad news and tough questions. However, our students' reactions should instead perhaps be a source of encouragement and reassurance. Our students' reactions show that they are engaged in the subject and have come to care about these issues. Furthermore, all friction and all tensions should not be avoided. It could be construed as our duty as teachers to bring them forth, as they are the precursors of change, and a necessary precondition for understanding where there is more work to do, where our efforts can have reverberations, and hence where we should direct our energies.

Note

1 www.wwoof.net/

References

Bardi, U. (2014). *Extracted: How the quest for mineral wealth is plundering the planet.* Chelsea Green Publishing.

Baumer, E. P., & Silberman, M. (2011). When the implication is not to design (technology). In *Proceedings of the SIGCHI Conference on Human Factors in Computing Systems* (pp. 2271–2274). New York: ACM. https://doi.org/10.1145/1978942.1979275

Bell, G., & Dourish, P. (2007). Yesterday's tomorrows: Notes on ubiquitous computing's dominant vision. *Personal and Ubiquitous Computing, 11*(2), 133–143.

Bidwell, N. J., Siya, M., Marsden, G., Tucker, W. D., Tshemese, M., Gaven, N., ... Eglinton, K. A. (2008, September). Walking and the social life of solar charging in rural africa. *ACM Trans. Comput.-Hum. Interact., 20*(4), 22:1–22:33. Retrieved from http://doi.acm.org/10.1145/2493524 doi:10.1145/2493524

Brynjarsdottir, H., Håkansson, M., Pierce, J., Baumer, E., DiSalvo, C., & Sengers, P. (2012, May). Sustainably unpersuaded: how persuasion narrows our vision of sustainability. In *Proceedings of the SIGCHI Conference on Human Factors in Computing Systems* (pp. 947–956). New York: ACM. http://dx.doi.org/10.1145/2207676.2208539

Cramer, B. W. (2012). Man's need or man's greed: The human rights ramifications of green ICTs. *Telematics and Informatics, 29*(4), 337–347.

Daly, H. E. (1996). *Beyond growth: The economics of sustainable development.* Beacon Press.
Dennis, K., & Urry, J. (2009). *After the car.* Polity.
Dobson, A. (2007). *Green political thought* (4th ed.). London: Routledge.
Eriksson, E., & Pargman, D. (2014). ICT4S reaching out: Making sustainability relevant in higher education. In *Proceedings of the International Conference ICT for Sustainability 2014 (ICT4S-14).* Atlantis Press.
Fry, T. (2007). Redirective practice: An elaboration. *Design Philosophy Papers, 5*(1), 5–20.
Füssel, H.-M., & Jol, A. (2012). *Climate change, impacts and vulnerability in Europe 2012* (EEA Report No. 12/2012).
Giddens, A. (2009). *The politics of climate change.* Cambridge, UK: Cambridge University Press.
Greer, J. M. (2008). *The long descent: A user's guide to the end of the industrial age.* New Society Publishers.
Greer, J. M. (2013). *Not the future we ordered: Peak oil, psychology, and the myth of progress.* Karnac Books.
Huesemann, M., & Huesemann, J. (2011). *Techno-fix: Why technology won't save us or the environment.* New Society Publishers.
Jackson, T. (2009). *Prosperity without growth: Economics for a finite planet.* Earthscan.
Knowles, B., & Eriksson, E. (2015). Deviant and guilt-ridden: Computing within psychological limits. *First Monday, 20*(8).
Levin, K., Cashore, B., Bernstein, S., & Auld, G. (2012). Overcoming the tragedy of super wicked problems: Constraining our future selves to ameliorate global climate change. *Policy Sciences, 45*(2), 123–152.
Lovelock, J. (2010). *The vanishing face of Gaia: A final warning.* Basic Books.
MacKay, D. J. (2009). *Sustainable energy – without the hot air.* Cambridge, UK: UIT.
Mann, S. (2011). *The green graduate: Educating every student as a sustainable practitioner.* New Zealand: New Zealand Council for Educational Research.
Mann, S., Muller, L., Davis, J., Roda, C., & Young, A. (2010). Computing and sustainability: Evaluating resources for educators. *ACM SIGCSE Bulletin, 41*(4), 144–155.
Meadows, D. H., Meadows, D. L., Randers, J., & Behrens, W. W. (1972). *The limits to growth.* Universe Books.
Meyerson, D. E., & Scully, M. A. (1995). Crossroads tempered radicalism and the politics of ambivalence and change. *Organization Science, 6*(5), 585–600.
Monbiot, G. (2007). *Heat: How we can stop the planet burning.* Penguin, UK.
Neves, L., Krajewski, J., Jung, P., & Bockemuehl, M. (2012). GeSI SMARTer 2020: The Role of ICT in Driving a Sustainable Future. *Global e-sustainibility initiative (GeSI).*
Norgaard, K. M. (2011). *Living in denial: Climate change, emotions, and everyday life.* Cambridge, MA: MIT Press.
Pargman, D., Ahlsén, E., & Engelbert, C. (2016). Designing for sustainability: Breakthrough or suboptimisation? In *Proceedings of the International Conference ICT for Sustainability 2016 (ICT4S-16).* Atlantis Press.
Pargman, D., & Eriksson, E. (2013). "It's not fair!": Making students engage in sustainability. In *Proceedings of Engineering Education for Sustainable Development.*
Pargman, D., & Eriksson, E. (2016). At odds with a worldview: Teaching limits at a technical university. *Interactions, 23*(6), 36–39.
Preist, C., Schien, D., & Blevis, E. (2016). Understanding and mitigating the effects of device and cloud service design decisions on the environmental footprint of digital infrastructure. In *Proceedings of the 2016 CHI Conference on Human Factors in Computing Systems* (pp. 1324–1337). New York: ACM.

Randall, R. (2009). Loss and climate change: The cost of parallel narratives. *Ecopsychology*, *1*(3), 118–129.

Rittel, H. W., & Webber, M. M. (1973). Dilemmas in a general theory of planning. *Policy sciences*, *4*(2), 155–169.

Shirky, C. (2008). *Here comes everybody: The power of organizing without organizations*. Penguin.

Silberman, M. S. (2015). Information systems for the age of consequences. *First Monday*, *20*(8).

Silberman, M., Nathan, L., Knowles, B., Bendor, R., Clear, A., Håkansson, M., . . . Mankoff, J. (2014). Next steps for sustainable HCI. *Interactions*, *21*(5), 66–69.

SpaceX. (2013). *Hyperloop white oaper*. Retrieved from www.spacex.com/sites/spacex/files/hyperloop_alpha-20130812.pdf

Steffen, W., Crutzen, P. J., & McNeill, J. R. (2007). The Anthropocene: Are humans now overwhelming the great forces of nature. *Ambio*, *36*(8), 614–621.

Steffen, W., Richardson, K., Rockström, J., Cornell, S. E., Fetzer, I., Bennett, E. M., . . . de Wit, C. A. (2015). Planetary boundaries: Guiding human development on a changing planet. *Science*, *347*(6223), 1259855.

Sterling, S. (2004). Higher education, sustainability, and the role of systemic learning. In P. B. Corcoran & A. E. J. Wals (Eds.), *Higher education and the challenge of sustainability: Problematics, promise and practice* (pp. 49–70). The Netherlands: Springer.

Stocker, T., Qin, D., Plattner, G., Tignor, M., Allen, S., Boschung, J., . . . Midgley, P. (2013). Climate change 2013: The physical science basis. *Working Group I Contribution to the Fifth Assessment Report of the Intergovernmental Panel on Climate Change. Summary for Policymakers*.

Tainter, J. (1990). *The collapse of complex societies*. Cambridge: Cambridge University Press.

Terranova, T., & Donovan, J. (2013). Occupy social networks: The paradoxes of corporate social media for networked social movements. In *Unlike Us Reader: Social Media Monopolies and Their Alternatives* (pp. 296–311). Institute of Network Cultures.

Weber, E. U. (2010). What shapes perceptions of climate change? *Wiley Interdisciplinary Reviews: Climate Change*, *1*(3), 332–342.

Weiser, M. (1991). The computer for the 21st century. *Scientific American*, *265*(3), 94–104.

Zerubavel, E. (2006). *The elephant in the room: Silence and denial in everyday life*. Oxford: Oxford University Press.

10 Participation in design for sustainability

Janet Davis and Sandra Burri Gram-Hansen

Introduction

A perennial conversation in interaction design is who has the power to make design decisions: The paying customer? The designers themselves? What role do future users play? These questions are particularly salient in design for sustainability. The effects of climate change are both global and local. Individual actions have little impact, yet organizational and societal changes require the political support and active participation of individuals. Everyone has a stake – whether they believe it or not.

Persuasive technology (PT) has been a dominant approach to sustainable HCI (DiSalvo et al., 2010). While many interactive technologies serve as tools to facilitate particular tasks – for example, spreadsheets organize and perform repetitive calculations – persuasive technologies are distinguished by the deliberate intention to change users' attitudes and behaviors (Fogg, 1998). For example, persuasive technologies in the realm of healthcare seek to increase physical activity through feedback and rewards, or train users to take medications regularly through reminders and praise. In the realm of sustainability, persuasive technologies seek to influence behaviors such as recycling, electricity use, and transportation mode choice. Because information technologies are persistent and omnipresent, Fogg (2003) argues they can influence behavior more effectively than a human persuader can.

Yet, the persuasive technology approach to sustainable HCI has been critiqued as oversimplifying not just the complexities of climate change, but also the complexities of people's lives. We can address both these critiques through participatory design. By engaging future users as partners in the design process, designers gain their expertise regarding the context in which the technology will be used. Moreover, engaging sustainability experts can overcome designers' limited understandings of sustainability and ground choices about design goals with assessment of real-world impacts.

In this chapter, we extend Fogg's understanding of persuasion with ideas from classical rhetoric and contemporary social psychology. We argue that participatory design is not just nice to have, but necessary. Participatory design is nice to have because it is an opportunity for learning that begins the process of behavior change. Participatory design is necessary because design for sustainability requires others' expertise: not only future users, but also sustainability experts. Finally, we discuss three recent cases of participatory design for sustainability, drawing out evidence in support of our claims.

Persuasive technology for sustainability

To understand sustainable HCI, it is important to understand the role of persuasive technologies. Persuasive technologies (PT) are defined as technologies designed with the intention to change user attitudes and behaviors (Fogg, 2003). PT is a predominant approach to sustainable HCI: DiSalvo, Sengers, and Brynjarsdóttir found in their 2010 review that 45% percent of sustainable HCI publications proposed persuasive technologies.

Persuasive technologies for sustainability typically seek to convince users to make more sustainable choices in everyday life. Some are explicitly educational and promote specific behaviors. For example, the PowerHouse computer game uses operant conditioning and praise to teach users everyday behaviors that save energy in the home, for example, cooking with the microwave instead of the conventional oven (Bång et al., 2006). Others provide feedback on behavior, and may or may not suggest sustainability as a motivation. For example, WaterBot provides feedback on water use at a shared sink and lets users compare their use with others' use (Arroyo et al., 2005). The intention to promote water conservation is not made explicit to the user. In a similar vein, UbiGreen aims to motivate low-impact transportation choices, like walking, biking, and public transit, by providing feedback on behavior. But by contrast, feedback is visualized as a growing tree or a family of polar bears – imagery typically associated with environmental sustainability (Froehlich et al., 2009).

Most of these technologies fall into the category of *behavior change support systems*, that is, technologies that individuals adopt to support a desired behavior change (Oinas-Kukkonen & Harjumaa, 2009). For example, UbiGreen is designed for users who already value environmental sustainability and want to change their transportation behavior. The problem with focusing on users who already want to change their behavior is that this approach is limited in scope. Most people are not just uninformed about sustainable practices; they are also unmotivated to change their behavior (Stoknes, 2015). WaterBot does not have this limitation, as it engages anyone using the shared sink, but it is also unclear whether WaterBot would have any lasting effect on users' attitudes or behaviors.

PowerHouse differs from WaterBot and UbiGreen in that it is explicitly educational in its intent. It seeks to teach new behaviors which the user can rehearse in simulation. From this perspective, persuasion is much like teaching: both take time, and both are facilitated by a desire to learn. In developing ethical principles for persuasive technology, Spahn (2012) argues that PT should educate rather than manipulate, and "ideally the aim of persuasion should be to end the persuasion" because the subject has changed their mind and persuasion is no longer needed. Social psychologists, too, view persuasion as an ongoing process during which a person is influenced by one or more strategic initiatives instigated by either a system or a person (Miller, 2002). In classical rhetoric, Aristotle draws a clear distinction between persuasion (persuasio) and manipulation (peithananke), where the latter is referred to as lies embellished to appear truthful, forceful enticement, or simply chocolate-covered broccoli. Rather, to persuade is to convince a person by use of reasonable arguments. Persuasion comprises an interaction between the persuader and the persuadee, resulting in the establishment of shared goals and agreement on an appropriate approach to achieving those goals (Hasle & Christensen, 2007; Fafner, 1997).

Each of these views – ethics, social psychology, and classical rhetoric – reinforces the idea that sustainable behavior change is based on a change in attitude. In order to reach such a change of attitude, the user must have processed new knowledge and reached a new understanding of the given context (Gram-Hansen & Ryberg, 2015). Each of the example technologies described above seeks to provoke users to reflect on their own behavior, whether in the simulated environment of the PowerHouse game, through the abstract feedback of the Waterbot, or through the environmentally themed feedback of UbiGreen.

But how impactful is PT for sustainability? In their 2010 review, DiSalvo et al. (2010) found that most work on PT for sustainability has been motivated by the development of PT theories or methods, with sustainability as the application. Target behaviors are chosen by PT designers who are not experts on sustainability. Evaluation typically focuses on usability or short-term behavior change, and not on attitude change or environmental impacts.

In their subsequent critique, Brynjarsdóttir et al. (2012) tease out further implications of this modernist approach. They argue that current approaches to sustainable HCI grant "undue authority" to designers, who are typically not experts on sustainability. Consequently, designs reflect a narrow view of sustainability as resource conservation, seeking to change individual behavior rather than systems and infrastructures. Even within that frame, only a few PTs, like PowerHouse, teach specific conservation behaviors. Many more, like UbiGreen and WaterBot, give quantitative feedback or promote "awareness". Few systems suggest appropriate goals for behavior change. Brynjarsdóttir et al., also characterize PT for sustainability as "too distant from lived use", abstracting away the constraints imposed on individual behavior by power relationships, such as those between tenants and landlords, and by social infrastructures, such as conflicting demands on working parents' time. PTs that are designed for an abstracted, universal user inevitably overlook values and concerns that are important to particular users' choices in a particular context of use.

Yet, context has always been an important part of persuasion, through the rhetorical notion of *Kairos*. In some of the earliest theoretical work on persuasive technology, Fogg (2003) defines Kairos as the opportune time to intervene. Finding this moment does require sensitivity to the user's context. But in classical rhetoric, Kairos is defined more comprehensively as a three-dimensional concept combining the appropriate time, place, and manner for a persuasive initiative to be effective (Kinneavy, 2002). While both *time* and *place* may potentially be identified by a qualified designer, the appropriate manner is dependent on users' perception of the context. For instance, a mobile app may be designed to provide the user with cues or notifications at opportune moments in time, but the appropriate manner for the app to prompt the user, with sound, vibration, or a visual notification, may depend on who the user is with and what they are doing. The content of a persuasive message, too, might be appropriate in some situations and frustrating or insensitive in others. Only those with direct experience of the use context can provide these understandings, underpinning the importance of user partnership in design.

Participatory design and persuasive technology

Because the future users of PT for sustainability provide irreplaceable insight into *Kairos*, they should be included as partners in the design process. Participatory design provides a long tradition of such engagement.

Emerging from an explicitly political context as part of the Scandinavian workplace democracy movement, participatory design has historically emphasized democratic values and empowerment of those who will be most affected by new technologies (Bødker et al., 1993). The most central perspective in participatory design is an unshakable commitment to ensuring that those who are to apply a given technology are actively involved in its design. As such, participatory design research is driven by an ongoing effort to develop collaborative design processes that ensure all participants – both users and designers – are equally acknowledged as experts in their own right (Simonsen & Robertson, 2012). This acknowledgement of different types of expertise can be referred to as engaging in the process with a *participatory mindset*.

According to Simonsen and Robertson, participatory design may be defined as:

> A process of investigating, understanding, reflecting upon, establishing, developing and supporting mutual learning between multiple participants in collective reflection-in-action. The participants typically undertake the two principal roles of users and designers where the designers strive to learn the realities of the user's situation while the users strive to articulate their desired aims and learn appropriate technological means to obtain them.
>
> (Simonsen & Robertson, 2012)

As such, participatory design traditionally constitutes a type of dialectic collaboration, in which design experts and domain experts collaborate in generating shared understandings of the context and of potential solutions.

In participatory design for the workplace, users contribute expertise beyond knowledge of their everyday activities and context. Users are also experts on the knowledge, ideals, and best practices of their field, and contribute to an understanding of how technology can enhance their skills and improve the quality of their product (Bødker et al., 1993). Drawing on the words of Tohidi et al. (2006), the latter expertise provides important insight into "getting the right design", while the former expertise is needed to "get the design right".

As participatory design has moved beyond the workplace, and particularly into educational settings, expertise is distributed across more stakeholders. For example, Scaife et al. (1997) applied participatory design methods to educational multimedia for teaching ecology to schoolchildren. Teachers are experts on pedagogy and on the material to be learned, while students are experts on their own lived experience. While teachers draw on their expertise to identify difficult topics and suggest pedagogical strategies, students serve as *domain experts* with their own insight into the learning process and the classroom environment. Teachers and students together evaluate new technologies in the classroom, each providing a distinct perspective. Similarly, Taxén (2004) engaged two distinct groups of stakeholders in participatory design of museum exhibits. Museum staff serve as *subject-matter experts* who guide the selection of topics and goals for exhibit design, and who provide accurate content. Visitors provide valuable insights based on their experiences interacting with museum exhibits. Such insights include design criteria (what makes exhibits "efficient, educational, and fun?"), concepts for new exhibits on a given topic, and formative evaluation of exhibit design.

We have already argued that effective PT for sustainability must educate the user. Thus, we see an analogous need to engage at least two different categories of stakeholders, moving from the traditional dialectic of participatory design to a new trialectic. First, we agree with Brynjarsdóttir et al. (2012), that design of PT for sustainability should include future users, who provide insight into their own lived experiences. In particular, representative users or domain experts provide insight into *Kairos*, which we defined earlier as the appropriate time, place, and manner for effective persuasion. But most users, like most PT designers, have a limited view of sustainability focused on individual consumption (Brynjarsdóttir et al., 2012; Knowles & Davis, 2016). Therefore, participatory design needs to also engage subject-matter experts, specifically sustainability experts, who can inform the adoption of design goals and approaches with potential for meaningful impacts. In short, sustainability experts can help us design the right thing, while representative users can help us design the thing right.

Finally, participatory design is itself an opportunity for persuasive engagement. At its best, participatory design is an extended process in which participants and designers learn from each other, develop shared understandings, establish common goals, and agree on approaches for achieving those goals. This fits closely with the concept of ideal persuasion developed in classical rhetoric and introduced in the previous section. As such, participatory design can be an initial step towards behavior change, as the design process affects the knowledge and attitudes of participants as well as designers.

Case studies

Next, we discuss three case studies in which participatory design approaches are applied to PT for sustainability:

- Participatory design of the Firefly Staircase, an ambient technology for a college campus;

- Participatory design with a college EcoHouse;
- Participatory design of sustainability training for conscripts to the Danish military.

There are two main reasons why we chose these case studies. First, there are few reported studies applying participatory design to sustainable HCI. Others that we are aware of (e.g., Prost et al., 2015) seek to understand technology users' attitudes towards sustainability, but do not engage in design of new technology. These are the only three reported studies we know of that apply participatory design approaches to designing PT for sustainability. Second, we have direct experience with these case studies. While all three have been previously reported in the literature, we can provide additional detail to support our argument.

For each case study, we briefly present the context, goals, and outcomes of the participatory design process. We discuss the expertise contributed by participants with respect to both *Kairos* and sustainability, as well as any gaps in expertise. Finally, we consider how the design process itself influenced the attitudes of participants.

Case study: the Firefly Staircase

Our first case study concerns participatory design on a college campus (Miller et al., 2009). The authors sought to design ambient persuasive technology for resource conservation in student spaces. Their rationale for using participatory design was to ensure the intervention was situationally appropriate and "owned" by members of the community.

The research team advertised for student volunteers – representative users – to participate in a series of design workshops. The team also consulted regularly with the campus sustainability coordinator – a subject-matter expert – who provided insight into sustainability impacts of possible interventions.

The first phase of the design process focused on selecting a site for intervention. Participants explored the campus looking for resource conservation opportunities, particularly noting where they saw other students engaged in unsustainable behavior. Between workshops, the research team met with the campus sustainability coordinator to gain insight into the environmental impacts of these behaviors.

The second phase focused on mocking up an ambient display. The researchers facilitated workshops in which participants generated and gave feedback on design concepts. The researchers also continued to consult with the campus sustainability consultant, not just on potential impacts but on the feasibility of data collection and system installation.

Ultimately, the site selected for the intervention included an overused freight elevator and a nearby staircase. The design goal was to encourage people to use the stairs rather than the elevator, which is better for the environment and for individual health. However, participants noted situations where people need to use the elevator: for example, if their mobility is impaired, or if they are moving heavy objects. Hence, participants deemed it inappropriate to chastise people for using the elevator. Instead, the group designed the Firefly Staircase, an interactive sculpture intended to make the stairs more attractive and show it can be faster to use the stairs than the elevator.

In this case study, the campus sustainability coordinator played an important role as subject-matter expert, guiding the research team towards impactful interventions. Student participants not only predicted their own responses, but also used empathy informed by their experiences with others in the community to anticipate likely responses to interventions and possible impacts on behavior and well-being. Thus, participants served as domain experts providing valuable insight into *Kairos*.

As intended, some participants became quite invested in the design and supported a permanent installation of the Firefly Staircase. While there was no attempt to assess the impact of participation on attitudes towards sustainability, the authors note that most participants were attracted to the project through friendships or an interest in design, not by any particular interest in sustainability. Thus, participants likely looked at sustainability in new ways.

Case study: EcoHouse

Many American colleges and universities have project or interest houses, where a small group of undergraduates with a shared goal or interest share a home. Grinnell College's EcoHouse is a group of 10 students living together with shared goals of living sustainably and educating other students about sustainability.

Davis (2010, 2012) carried out an open-ended participatory design study with EcoHouse residents over an academic year. The campus sustainability coordinator had a smaller role in this project than in the previous project, focused mainly on access to existing information technology. Rather, EcoHouse residents each brought their own expertise in particular areas such as energy conservation, environmental health, local foods, and composting. Through exploratory activities including document review, cultural probes, and participant-observation, the facilitator learned about the values and rhythms of EcoHouse, while residents engaged in reflection and creative expression. Exploratory activities also helped to share power by meeting residents on their ground and to establish respect for residents' expertise.

Concept generation workshops (Davis, 2010) were the first design activity conducted in the facilitator's space, which served to distance participants from their lived experience. Participants took the lead in developing design concepts based on their own goals, activities, and values. Some discussion concerned not the usefulness or feasibility of design concepts, but rather their contextual appropriateness. The workshops were seeded with example technologies, such as a device attached to the showerhead which cuts off the flow of water at a preset time. While acknowledging this device would save water, residents thought it was inappropriate to install in EcoHouse and in other showers on campus. They reasoned that it was too coercive and would not educate users about the importance of conserving water and energy. Moreover, having an absolute cutoff on shower length would not account for situational differences

Participation in design for sustainability 173

like an individual's hair length, or strategic choices such as alternating frequent quick showers with occasional "luxury showers". Moreover, participants raised concerns about the paradox of installing new electronic devices to raise awareness of energy consumption.

Residents deliberated upon the design concepts and choose a few to pursue. The "props board" and "sustainability diary" were low-technology strategies for encouragement and reflection, respectively, which EcoHouse residents implemented on their own shortly after the workshops. The most successful initiative was a joint exploration of devices for preventing phantom load. The facilitator purchased several off-the-shelf devices; EcoHouse residents experimented with using them in different locations around the house. They reflected on which devices were easiest to use and most likely to result in energy conservation rather than waste when used by students. Residents finally made recommendations for which devices to adopt and how to best use them.

In this study, EcoHouse residents served as both domain experts and subject-matter experts, with mixed results. As domain experts, participants provided valuable insight into *Kairos* in a range of use contexts on campus, as well as EcoHouse itself. Participants were able to provide some subject-matter expertise and were keen to continue learning; closer engagement with more senior subject-matter experts such as the campus sustainability coordinator would have facilitated such learning. Moreover, as some participants would be quick to point out, they have a conflict of interest. Their comfort was often at odds with what their knowledge of sustainability told them was best to do. Having someone in the room to represent sustainability alone, without that conflicting interest, may have brought some more radical directions to consideration.

Finally, participants themselves observed that design workshops were an opportunity for participants to reflect and learn. Not every resident was able to attend the concept generation workshop, but one resident found the discussion so fruitful that he "wish[ed] everyone could have been [t]here".

Case study: the Danish military

Green transition in a complex organizational context is exemplified with the Danish military. The Danish military is a diverse organisation with a combination of armed forces, technical staff, and civilian employees. Employee tasks and work facilities differ greatly: 33% of military employees are office workers, while others rarely work indoors. Besides its role in international collaborations, the military is an educational institution, where young men and women are educated to become part of the Danish armed forces. Within the Danish military, eligible young men are drafted for four months' basic military training, after which 25% continue with further military education. The draft policy is intended to ensure diversity within the military and consequently ensure that the Danish military remains representative of Danish societal values.

This case addresses the environmental education of army conscripts. The Danish Ministry of Defense has presented an ambitious climate and energy strategy, which aims to lower energy consumption and minimize CO_2 emissions. Overall, the Danish military aims to:

- Lower energy consumption by 20% compared to 2006;
- Increase use of energy from sustainable sources by 60%;
- Reduce CO_2 waste by 40% compared to 1990.

Energy- and environment-related issues are not a core area for most employees in the Danish military. While many are positive towards the idea of an environmentally friendly workplace, some find that the strategic goals conflict with the primary tasks of the military. This reluctance is partly based on strong traditions and hierarchical structures within the military, and partly by practical challenges commonly seen in large educational institutions, where too many subjects must be covered in too little time.

In order to abide by Danish as well as EU regulation, all military employees are required to complete courses in sustainability-related subjects such as appropriate waste administration and disaster management. However, in practice, it proved challenging to include environmental education in the short time frame of basic training. As a result, initiatives have sought to produce digital learning resources which not only educate military employees in appropriate behavior in relation to sustainability, but also promote positive attitudes towards becoming a more green organization (Gram-Hansen & Ryberg, 2015).

Early designs were traditional e-Learning solutions, which require the user to sit by a computer and work through materials and quizzes. However, the vast majority of military employees spend little or no time in front of a computer during a normal work day. From a practical perspective, it was not feasible to make special arrangements for all employees to spend time in a computer lab. Moreover, it proved to challenging for employees to transfer the information provided through e-Learning into daily practice.

To address these challenges, the e-Learning approach was replaced by a persuasive learning design that enabled the military staff to engage in the learning experience during their day-to-day routines. Participatory design was a key aspect of this approach.

A work group was established consisting of technical experts, persuasive designers, and experts in the military domain. This work group was engaged throughout the entire design and implementation process. During this process, the transition from dialectic to trialectic roles of expertise became evident due to the challenges of designing for sustainability. While participatory design traditionally calls for collaboration between designers and domain experts, the case presented an unquestionable need to also include subject-matter experts, as neither designers nor domain experts were sufficiently knowledgeable about sustainability issues and potential. These include areas such as legislative requirements for environmental behavior, ministerial strategies for the organization, distinct challenges within the specific case (e.g., places of excessive water and energy usage, or incorrect waste management), as well as areas for improvement.

The participatory design approach was found to contribute in several ways. First, it facilitated a deeper understanding of the context, enabling the designers to consider the values and practices of the users in the new design solution. Moreover, the participatory approach also became a key element in the process of attitude change. Through participatory design, the domain experts gained ownership of the developed initiatives, leading the participants to not only identify them as applicable and relevant, but also resulting in a more positive attitude towards sustainability issues as part of the curriculum for basic military training.

Discussion

The case studies show how participatory design activities not only facilitated the designers' understanding of appropriate manner within the distinct context, but also motivated attitude change within the participants. From activities such as those described in the EcoHouse case, participants were prompted to engage in discussions about appropriate solutions for their situation, and also to discard solutions which they found coercive or counterproductive. Similarly, participants in the design of the Firefly Staircase were able to anticipate inappropriate impacts on those who cannot reasonably change their behavior. In both cases, participants took ownership of the solutions. In the case of the Danish military conscripts, the participatory design activities served as a trigger for a change towards a more positive attitude towards the green transition agenda amongst the involved army instructors. The workshops enabled them to influence and take ownership of the design solutions, and provided them with the opportunity to reflect upon and articulate why the subject was in fact highly relevant for the army conscripts.

As future users were needed to understand *Kairos*, subject-matter experts were needed to understand solutions for sustainability. In the Firefly Stairway case, the participants had an interest in sustainability, but no particular expertise. The designers needed to engage a subject-matter expert, the campus sustainability coordinator, to help identify opportunities for positive impact. Similarly, the military case exemplified how the involved domain experts (military instructors) did not hold sufficient knowledge about solutions for sustainability, nor did their attitude towards the subject motivate them to engage and participate in the design process. It was in the sharing of the power to design, where designers, domain experts, and subject-matter experts all contributed, that a mutual understanding of appropriate solutions was established.

Of the three cases, the EcoHouse case most closely aligned with the traditional dialectic of participatory design, in that participants contributed expertise on the subject matter of sustainability as well as expertise on their own practices. Even there, the designer engaged the campus sustainability coordinator to evaluate some of the proposed solutions for feasibility and impact. Paradoxically, the greatest successes of this project did not involve the design of new technology (Baumer & Silberman, 2011). The designer spent quite some time pursuing a project of great interest to some participants – helping them remember to turn off the oven – whereas a discussion of this problem with the campus sustainability coordinator would probably have led to non-design solutions such as buying a new oven with an automatic shut-off timer. Engaging in trialectic conversation amongst the designer, the campus sustainability coordinator, and the EcoHouse residents, would have led to greater learning overall.

Of course, the challenges of sustainability are not confined by organizational boundaries. Nor do sustainability experts agree on what is best to do in every situation, or even on what defines *sustainability*. These case studies are limited in that all three were conducted within well-defined organizations, with sustainability experts who view sustainability as primarily resource conservation. To precipitate change at a societal level, future work must move beyond organizational boundaries and beyond the notion of a single expert viewpoint on what is best for sustainability. For example, Hirsch and Anderson's (2010) qualitative study of water use in central New Mexico reveals a complex landscape of socially situated practices carried out by stakeholders with a wide range of goals, fears, and aspirations. There is not a single understanding of *sustainability* or even *consumption*. The science is difficult, and there is not a unified "expert" voice. Introducing persuasive technology into such a setting would require trust and mutual education.

Participatory design seems well suited to such a challenging political context. Indeed, Buhlmann (2014) proposes participatory design as a suitable approach for international rule-making on global sustainability concerns – an even more complex political problem with an even greater diversity of stakeholders. Including multiple subject-matter experts with differing expertise would move even further beyond the traditional dialectic to provide even greater opportunities for creative dissent and collaboration.

Conclusion

In this chapter, we have considered the role of persuasive technology in design for sustainability. To have an impact on environmental sustainability, a one-time change in behavior is not enough. Behavior change for sustainability must itself be sustained. We have argued that sustained behavior change involves a learning process which takes place over time. When future users of persuasive technology participate in the design process, that experience begins the process of learning, reflection, and attitude change.

We have further argued that designers benefit from partnership not only with future users, but also with subject-matter experts. Sustainability experts provide guidance with respect to appropriate goals for design, while future users provide key insights into *Kairos*, or the appropriate manner of persuasion. While each kind of expertise is necessary for effective design, neither is sufficient on their own. Hence, participatory design for sustainability should engage both subject-matter experts and future users, moving from the traditional dialectic of participatory design towards a new trialectic.

References

Arroyo, E., Bonanni, L., & Selker, T. (2005). Waterbot: Exploring feedback and persuasive techniques at the sink. In *Proceedings of the SIGCHI Conference on Human Factors in Computing Systems* (pp. 631–639). New York: ACM.

Bång, M.,Torstensson, C., & Katzeff, C. (2006). The PowerHouse: A persuasive computer game designed to raise awareness of domestic energy consumption. In *Proc of First International Conference on Persuasive Computing for Well-Being*.

Baumer, E. P. S., & Silberman, M. S. (2011). When the implication is not to design (technology). In *Proceedings of the SIGCHI Conference on Human Factors in Computing Systems* (pp. 2271–2274). New York: ACM.

Bødker, S., Christiansen, E., Ehn, P., Markussen, R., Mogensen, P., & Trigg, R. (1993). The AT-Project: Practical research in cooperative design. *DAIMI PB 454*. Aarhus, Denmark: Computer Science Department, Aarhus University.

Bødker, S., Grønbæk, K., & Kyng, M. (1993). Cooperative design: Techniques and experiences from the Scandinavian scene. In D. Schuler & A. Namioka (Eds.), *Participatory design: Principles and practices* (pp. 157–175). Hillsdale, NJ: Lawrence Erlbaum Associates.

Brynjarsdottir, H., Håkansson, M., Pierce, J., Baumer, E., DiSalvo, C., & Sengers, P. (2012). Sustainably unpersuaded: How persuasion narrows our vision of sustainability. In *Proceedings of the SIGCHI Conference on Human Factors in Computing Systems* (pp. 947–956). New York: ACM.

Buhlmann, K. (2014). Enhancing collaborative rule-making on global sustainability concerns through participatory design: A research agenda based empirically on United Nations developments on business conduct. In *Proceedings of the 13th Participatory Design Conference: Short Papers, Industry Cases, Workshop Descriptions, Doctoral Consortium papers, and Keynote abstracts – volume 2 (PDC '14)* (pp. 63–66). New York, NY: ACM.

Davis, J. (2010). Generating directions for persuasive technology design with the inspiration card workshop. In *Proceedings of the 5th International Conference on Persuasive Technology (PERSUASIVE '10). LNCS 6137* (pp. 262–273). Verlag, Berlin, Heidelberg: Springer.

Davis, J. (2012). Early experiences with participation in persuasive technology design. In *Proceedings of the 12th Biennial Participatory Design Conference (PDC '12)* (pp. 119–128). New York, NY: ACM.

DiSalvo, C., Sengers, P., & Brynjarsdóttir, H. (2010). Mapping the landscape of sustainable HCI. In *Proceedings of the SIGCHI Conference on Human Factors in Computing Systems* (pp. 1975–1984). New York: ACM.

Fafner, J. (1997). "Retorikkens brændpunkt." Rhetorica Scandinavica Maj 1997(nr. 2): 7–19

Fogg, B. (2003). *Persuasive technology, using computers to change what we think and do.* Morgan Kaufmann Publishers.

Fogg, B. J. (1998). Persuasive computers: perspectives and research directions. In *Proceedings of the SIGCHI conference on Human factors in computing systems* (pp. 225–232). ACM Press.

Froehlich, J., Dillahunt, T., Klasnja, P., Mankoff, J., Consolvo, S., Harrison, B., & Landay, J. A. (2009). UbiGreen: Investigating a mobile tool for tracking and supporting green transportation habits. In *Proceedings of the SIGCHI Conference on Human Factors in Computing Systems* (pp. 1043–1052). New York: ACM.

Gram-Hansen, S. B., & Ryberg, T. (2015). Attention – influencing communities of practice with persuasive learning designs. In *Proceedings of Persuasive Technology: Tenth International Conference* (pp. 184–195). Springer.

Hirsch, T., & Anderson, K. (2010). Cross currents: Water scarcity and sustainable HCI. In *Extended Abstracts on Human Factors in Computing Systems* (pp. 2843–2852). ACM.

Kinneavy, J. L. (2002). Kairos in classical and modern rhetorical theory. In P. Sipiora & J. S. Baumlin (Eds.), *Rhetoric and Kairos, essays in history, theory and practice*. State University of New York Press.

Kjær Christensen, A.-K., & Hasle, P. F. V. (2007). Classical rhetoric and a limit to persuasion. In *Proceedings of Persuasive Technology: Second International Conference*. Revised Selected Papers. Springer.

Knowles, B., & Davis, J. (2016). Is sustainability a special case for design? *Interacting with Computers, 29*(1), 58–70.

Miller, G. R. (2002). On being persuaded, some basic distinctions. In J. P. Dillard & M. Pfau (Eds.), *The persuasion handbook, developments in theory and practice*. London: Saga Publications.

Miller, T., Rich, P., & Davis, J. (2009). ADAPT: Audience design of ambient persuasive technology. In *Extended abstracts on human factors in computing systems* (pp. 4165–4170). New York: ACM.

Oinas-Kukkonen, H., & Harjumaa, M. (2009). Persuasive systems design: Key issues, process model, and system features. *Communications of the Association for Information Systems, 24*(1), 28.

Prost, S., Mattheiss, E., & Tscheligi, M. (2015). From awareness to empowerment: Using design fiction to explore paths towards a sustainable energy future. In *Proceedings of the 18th ACM Conference on Computer Supported Cooperative Work & Social Computing* (pp. 1649–1658). New York: ACM.

Simonsen, J., & Robertson, T. (2012). *Routledge International Handbook of Participatory Design*. Routledge.

Scaife, M., Rogers, Y., Aldrich, F., & Davies, M. (1997, March). Designing for or designing with? Informant design for interactive learning environments. In *Proceedings of the SIGCHI Conference on Human Factors in Computing Systems* (pp. 343–350). New York: ACM.

Spahn, A. (2012). And lead us (not) into persuasion. . .? Persuasive technology and the ethics of communication. *Science and Engineering Ethics, 18*, 633–650.

Stoknes, P. E. (2015). *What we think about when we try not to think about global warming: Toward a new psychology of climate action*. Chelsea Green Publishing.

Taxén, G. (2004). Introducing participatory design in museums. In *Proceedings of the Eighth Conference on Participatory Design: Artful Integration: Interweaving Media, Materials and Practices – Volume 1* (Vol. 1, pp. 204–213). New York: ACM.

Tohidi, M., Buxton, W., Baecker, R., & Sellen, A. (2006). Getting the right design and the design right. In *Proceedings of the SIGCHI Conference on Human Factors in Computing Systems* (pp. 1243–1252). New York: ACM.

Response 3a Connected and complicit

Mél Hogan

As an Environmental Media scholar, I was invited to respond to a series of chapters at the intersections of human-computer interaction (HCI) and sustainability – a topic I have broadly been thinking about in my own research, but also, and more intensively, grappling with in my teaching. Undergraduate students (in various fields intersecting with questions of data/media) are increasingly primed to ask for "real-life" "solutions" to environmental problems, and especially those that relate in some way to their digital lives and conveniences. Electronic waste (e-waste), the chain of production of devices and its labor conditions, energy consumption and the massive impacts of data storage and circulation, etc., are all themes that students have a reaction to. But by "solutions", students generally desire immediate, hands-on, tangible interventions, and almost always fail to see the importance and impact of critical theory, culture, and policy in the process. There is a sense that we live in a perpetual now, and that only technology can save us from the conditions (of capitalism) that have brought us to the brink. How, then, might sustainable HCI (SHCI) open up this issue and address the fact that we can no longer avert the impending ecological disaster without also ridding ourselves of its underpinning neoliberal logics?

Eriksson and Pargman address this very subject through a series of anecdotes used to reflect on the expectations and barriers they face when teaching sustainability to engineering students at a technical university. Like mine, their students call for optimism and a positive outlook on the future, not realizing that this is in itself a means of coping with the very real, very urgent demands of climate change. Drawing attention to "emotional barriers", Eriksson and Pargman bring to light the important (but often understated) issue of teaching sustainability in a place where the students demonstrate that they care about the future of the planet in part by rejecting the facts that outline a bleak outlook, and favoring instead the quick and clean tech-fix that they have likely been sold within the market-driven context in which they study. On the one hand, this outlook generates hope through denial, and, on the other, insists on generating new modalities for participation and persuasion for students and/as citizens. These new modalities will have to mean a shift from individual to corporate responsibility, and – most urgently – the forming of a newfound collective strategy.

The question of participation and persuasion is addressed by Davis and Gram-Hansen. Their chapter draws attention to the importance of collaboration for the design of sustainability apps and tools, bringing together experts in the psychological, technical, and conceptual facets of design, but also including users as partners in the process. In designing for sustainability, the authors focus on persuasive technology (PT), which differs from more common task-oriented applications that lead users to track, organize, or complete specific tasks. Instead, PT works with the intention of changing behaviors and attitudes. When PT is used to promote principles of sustainability, it focuses on nudging users to have a greater and ongoing awareness of their recycling practices, consumption habits, and modes of transportation, etc., by way of gamified self-tracking. However, because these tactics have a tendency to appeal to those already invested in repair and already motivated to change, PT must invite a larger user base to the design table. Findings from the three case studies explored demonstrate that users involved in design do become more aware of the problem, more attentive to shifting contexts, and more positive in their understanding of sustainability. Getting users involved in design, however, also raises policy questions and ethical concerns, such as: Who will be responsible for the unexpected impacts or uses? How will personal data be managed? And/or how will user participation, skills, and labor be valued, credited, or remunerated?

The role of stakeholders and policy concerns are addressed in the other two chapters – by Thomas, and by Remy and Huang – both of which look at issues of waste and obsolescence.

Thomas explains policy as the rules and objectives that guide specific governmental agencies – which themselves draw on existing laws and norms – and considers the role HCI experts could play in future policymaking that also addresses environmental concerns. Through a thorough investigation of the Waste Electrical and Electronic Equipment (WEEE) public policies and implementation, the author makes a case for the unfortunate disconnect between policy and practice, namely through the curtailing of the Basel Convention, which allows for "dumping grounds" in lieu of the promised recycling and refurbishing of equipment in good working order. Policies set in place do account for global disparities, and in this case, attempt to prevent the Global North from dumping onto the Global South. But more involvement from a diverse pool of HCI experts, among others, is required to ensure that these policies work, hold more ground, where breaches have greater consequences, and become less susceptible to the winds of economic – and increasingly technological – growth that so far require a place to discard the excesses resulting from the planned obsolescence of our devices and their logics.

Remy and Huang make a case for "real-world" and "useful" SHCI outputs to contribute to a sustainable future, reaching beyond the realm of academia. The authors discuss the gap between theory and practice in HCI by first showcasing an example that demonstrates the success of merging the two, and secondly, by highlighting issues relating to deciphering theory for implementation purposes. When scholars are able to bring their own theoretical framework into practice, the process can be more straightforward; however, it does less to elucidate the gap, or the processes involved in implementing theoretical concepts. While the authors do not focus mainly on the permeability of concepts, cultural commentary, or theories, their work provides an important applied strategy for dealing with complex networks of stakeholders. Their passion for "real-world impact" through HCI knowledge translation speaks directly to today's students' anxieties, that is, a desire to see something change in their lifetime and by their own means.

Overall, the chapters in this section come together to highlight many facets of HCI with a focus on sustainability and sustainable design. They bring to the fore the urgency of climate change and the ongoing destruction of the environment, connected – however awkwardly – to our addiction and obsession with technology. By highlighting rather than denying the complexities of our current media landscape, the authors each raise important points about the future of design by making all of us – as users and as citizens – involved in the processes to change the logics, infrastructures, policies, standards, and contexts that have led us to this current conundrum.

Response 3b From participatory design to participatory governance through sustainable HCI

Rónán Kennedy

The four chapters in this part deal with the interaction between HCI research and teaching on the one hand, and HCI policy and practice on the other. They illustrate issues of conflict, control, and communication that are at the heart of discourse in democratic societies and can help us to explore the increasing importance of information and communications technologies (ICT) in the day-to-day functioning of economic and social institutions, creating a space for reflection on the challenge and potential of integrating values and ethics into academic research projects, curricula, and external engagement.

Issues of *conflict* are explicit in Eriksson and Pargman's contribution: between the need to be honest with students but avoid distressing them, negotiating with colleagues about the very real changes to teaching required by a high-level commitment to sustainability, and the challenges for students of translating that internalised orientation into the commercial reality that awaits after graduation. They lie just under the surface of the accounts provided by Davis and Gram-Hansen, in the various perspectives of experts in different disciplines, but also in the very divergent approaches to the development and deployment of technological artifacts. These can be quick (but ultimately shallow) fixes, reflecting incomplete understandings of sustainability, and are often aimed more at creating opportunities for managerialist interventions.

This impulse, which often motivates approaches to problems which rely on digital technology, is ultimately rooted in the hope that ICT offers a mechanism for creating immutable mobiles (Latour,1986), allowing comprehensive and inescapable *control* of individuals, even at considerable distances of space and time. Projects such as PowerHouse, WaterBot, and UbiGreen can be connected to soft regulation approaches, inspired by Sunstein and Thaler's "Nudge" (Thaler & Sunstein, 2008), which can ultimately become assemblages of eco-governmentality (Malette, 2009). Davis and Gram-Hansen present us with an alternative in participatory design. This involves the users in the development from the outset, thus drawing on what James C. Scott would call "metis", or local and practical knowledge (Scott, 1998, pp. 311–313), to build systems that work with, rather than against, the natural inclinations of those whose participation is essential to the success of the system. The appropriateness of this approach is particularly obvious in the EcoHouse case study.

The issue of control also emerges strongly in Thomas' chapter, although here it is the HCI research community which seeks to assert control, bringing to bear its knowledge and expertise on the environmental harms which can be caused by Waste Electrical and Electronic Equipment (WEEE), and the opportunities for better outcomes which exist in Green Public Procurement (GPP) initiatives. Thomas argues that the HCI community has much to contribute to policy- and law-making in this problem domain, and that should assert itself more strongly to make its impact felt.

The approaches to control delineated in these chapters seem initially paradoxical: Davis and Gram-Hansen advocate less initial control by designers in the initial stages of a project in order to produce an outcome with more legitimacy and context-appropriateness, which is therefore more likely to achieve and sustain leverage over individual behavior in the long run (even where, as in the Eco-House project, it does not involve new technology). Thomas points out that SHCI research will attain its greatest impact the more it strives to operate outside its traditional domain.

These dualities of conflict and control can be resolved by considering carefully how the SHCI field *communicates*: Remy and Huang consider in detail whether there are more appropriate or effective ways to speak to policy- or law-makers, colleagues in other academic disciplines, practitioners, or students? Thomas highlights issues of engaging at different scales of policy, but considering her contribution in tandem with Remy and Huang's brings to the fore the necessity of always considering the different discourses – technical, regulatory, economic – that are intertwined in the lengthy processes of design and decision-making which culminate in a functioning artifact.

The tone of the four chapters is not optimistic, each stressing the practical challenges, missed opportunities, and existing failures that confront a SHCI researcher who aims to make a difference through his or her work. Nonetheless, taken together, they also offer a positive foundation which can enable the discipline to engage with external interest groups in positive and constructive ways. The importance of ICT in modern society is undeniable. Applying actor-network theory, or at least a particular iteration of it (Latour, 1999), we can think of the devices which emerge from HCI design processes as "rhizomes", spreading versions of so-called sustainability practices in social, regulatory, and commercial contexts. What emerges clearly from these chapters is that the vision of sustainability which is embedded and embodied in these artifacts and assemblages is often incomplete, contested, and sometimes even destructive. However, the chapters also offer an alternative approach for the discipline, one in which what Eriksson and Pargam call "strong sustainability" approaches to teaching, participatory design, and external engagement can begin to challenge these unfortunate tendencies. Participatory design, in particular, can be linked to participatory governance methods (Paquet, 2001; Asaro, 2000), highlighting the importance of ethics and values in the designs of information artifacts and information systems (Kesan & Shah, 2004) in ways which enable ordinary users to have a voice and means to be heard (Stahl, 2011). In this way, SHCI can help to re-assert democratic control over the information infrastructures (Hanseth & Monteiro, 1998) which play an increasingly important role in all of our daily lives and thus avoid the "unproblematic techno-fix".

References

Asaro, P. M. (2000). Transforming society by transforming technology: The science and politics of participatory design. *Accounting, Management and Information Technologies, 10*(4), 257–290.

Hanseth, O., & Monteiro, E. (1998). *Understanding information infrastructure*. Retrieved August 27, 2014, from http://heim.ifi.uio.no/oleha/Publications/bok.pdf

Kesan, J., & Shah, R. (2004). Deconstructing code. *Yale Journal of Law and Technology, 6*, 277–389.

Latour, B. (1986). Visualisation and cognition: Drawing things together. In H. Kuklick (Ed.), *Knowledge and society studies in the sociology of culture past and present* (pp. 1–40). Greenwich: Jai Press.

Latour, B. (1999). On recalling ANT. In J. Law & J. Hassard (Eds.), *Actor network theory and after* (pp. 15–15). Chichester: Wiley.

Malette, S. (2009). Foucault for the next century: Eco-governmentality. In S. Binkley & J. Capetillo (Eds.), *A Foucault for the 21st century: Governmentality, biopolitics and discipline in the new millennium* (pp. 221–239). Newcastle upon Tyne: Cambridge Scholars Publishing.

Paquet, G. (2001). Smart communities and the geo-governance of social learning. *Optimum Online, 31*(2), 33–50.

Scott, J. C. 1998. *Seeing like a state: How certain schemes to improve the human condition have failed*. Yale University Press: New Haven.

Stahl, B. C. (2011). What future? Which technology? On the problem of describing relevant futures. In M. Chiasson, O. Henfridsson, H. Karsten, & J. I. DeGross (Eds.), *Researching the future in information systems* (pp. 95–108). Berlin: Springer.

Thaler, R. H., & Sunstein, C. R. (2008). *Nudge: Improving decisions about health, wealth, and happiness*. New Haven, CT: Yale University Press.

Photo Essay 5

Airstream traveling home (2011). A modern version of the iconic Airstream originally designed in the 1930s illustrates in some way the combination of luxury with the spirit of living small. Reflection on: Desjardins: *Reflections on longevity, unfinishedness and design-in-living.*

Eli Blevis

Part 4
Inspiring futures

11 A sustainable place

Everyday designers as placemakers

Audrey Desjardins, Xiaolan Wang, and Ron Wakkary

Introduction

Place is not only a geographic concept. In a place, we recognize and find familiar ensembles of artifacts, people, and spaces through spatial experiences and senses, perception, and movement (Tuan, 1977). It is *"a world that is not only perceived or conceived but also actively lived and receptively experienced"* (Casey, 2001). Relevant to place, the concept of placemaking – as originally developed by Jane Jacobs and William Whyte – focuses on the ongoing, collaborative, and incremental process of creating and sustaining a place. In this chapter, we investigate how everyday designers can act as placemakers and how this new perspective can further advance our understanding of sustainability in human-computer interaction (HCI).

Among various approaches to sustainable HCI, one perspective is to consider the "user" as an active producer and maker of his or her own artifacts and systems. This new identity reveals users as everyday designers who "together create and redesign artifacts long after the products have left the hands of professional designers" (Wakkary & Tanenbaum, 2009, p. 365). This new identity, the authors argue, supports sustainability through concepts of longevity, reinvention, and reuse, principles previously proposed by Blevis in his seminal paper on sustainable interaction design (Blevis, 2007). More broadly, HCI researchers have addressed sustainability and user empowerment and agency in previous research (e.g., (Kim & Paulos, 2011; Roedl et al., 2015; Wakkary et al., 2013; Woodruff et al., 2008)). This research often takes a close focus on artifacts and rarely addresses how everyday designers create places. In this chapter, we turn to the idea of placemaking as a way to investigate how everyday designers not only appropriate and transform artifacts but also how they are engaged in the making of a whole environment. This is a timely topic that can offer a contrasting perspective to topics like ubiquitous computing, home automation, and the Internet of Things. As technology is becoming more embedded in various aspects of our environments, it is necessary to also ask how everyday people might engage incrementally and over time with these environments. We are particularly concerned with investigating how everyday designers as placemakers can offer a new perspective on sustainable HCI and interaction design.

We explore how placemaking takes form in two cases: an autobiographical study of the making of a camper van interior and an ethnographic study of community gardeners in the city of Vancouver, Canada. The camper van represents a place for its inhabitants regardless of where it is located. The community garden is a place for its gardeners mostly for the connections and routines it encourages within the neighborhood. In the following sections, we start by offering a brief overview of sustainable placemaking as a motivation for the rest of the chapter. Next, we present literature that addresses sustainable interaction design through individual behavior, the empowered individual, and social and collective design. We continue by presenting in detail the two case studies: the conversion of a cargo van into a camper van and the ongoing making of a neighborhood community garden. We focus on the themes of *longevity*, *unfinishedness*, and *multiplicity* that emerged from our analysis of the case studies. We conclude with a discussion on sustainable interaction design and placemaking.

Background: sustainable placemaking

Placemaking was originally developed in the 1960s by Jane Jacobs and William Whyte, who proposed a place-based community-centered approach to urban planning (Project for Public Spaces, 2017). Their vision for urban planning and landscape architecture embraces community-driven initiatives, collaboration, inclusivity, transformation and adaptation over time, and a clear relation to physical and historical context. Since the 1960s, urban planners, landscape architects, architects, and philosophers have pushed further the concept of placemaking to also include "sustainable placemaking". While our focus in this chapter is not placemaking at the scale of urban planning, we found many aspects of sustainable placemaking to be highly relevant (and inspiring) for our discussion of the connections between everyday designers and placemaking. Below, we describe further the aspects of identity, community, long-term thinking, and a holistic approach – all aspects that are central to sustainable placemaking and that are relevant to our two cases.

Sustainable placemaking is community-oriented and aims at building a sense of belonging and identity (Heller & Adams, 2009). As Schneekloth and Shibley put it, "The making of places – our homes, our neighborhoods, our places of work and play – not only changes and maintains the physical world of living; it also is a way we make our communities and connect with other people" (Schneekloth & Shibley, 1995, p. 1). This sense of belonging and identity supports the development and maintenance of a healthy community and promotes well-being amongst its participants. While being community-oriented and often grassroots, sustainable placemaking also relies on a productive relationship between community, policy, and governing entities since places are oftentimes within the public realms of cities (Franklin & Marsden, 2015). Current debates explore how community-led sustainable initiatives can be less disconnected from government policy, and, in the reverse, how government-led initiatives can also become closer to community projects and visions (Franklin & Marsden, 2015). This aspect is highly relevant

A sustainable place 191

in the planning and design of cities, and, as we will describe, in our community gardens case.

Moreover, the sustainable aspect of placemaking is often referred to as the making of a place for today's generation as well as future generations (Potter, 2009), highlighting the importance of longevity and the temporal dimension of placemaking. Specifically, Potter (2009) describes the tension between a long temporal focus of sustainability and the ephemeral quality of the ongoing processes of design. Finally, sustainable placemaking proposes a holistic approach to sustainability, including discussions around economic, environmental, and social issues (Heller & Adams, 2009). Heller and Adams (2009) discuss how a multiplicity of approaches and disciplines are necessary to successfully create a place that allows a community to function and be sustainable.

Our goal in this chapter is to utilize sustainable placemaking in ways that are "scaled down" from urban planning to further our understanding of the relations between the everyday designer and his or her approach of making a place in order to contribute to the discussion of the importance of user agency to sustainability. We explore what insights a scaled version of placemaking could bring to current discussions of interaction design and HCI perspectives in sustainability. In particular, we use our two cases (the camper van and the community gardens) as a way to refine our understanding of sustainable placemaking in a way that is insightful for HCI and interaction design.

Related works: sustainable interaction design, everyday design, and social design

In this section, we present a short overview of changes that have occurred in recent research in sustainable interaction design. Importantly, we also aim at highlighting how various perspectives in sustainable interaction design support – or don't support – discussions around the idea of placemaking. More precisely, we offer a look at how sustainable interaction design perspectives moved from focusing on changing individual behavior, to supporting the "empowered individual" towards modifying his or her own place, to broader discussions around collective action to contend with social issues.

Sustainable interaction design

In HCI, "sustainability" as a lens to design technology was proposed by Blevis in 2007. To encourage environmental sustainability, Blevis suggests that interactive systems could be designed to promote individuals' sustainable behaviors in regard to the use of resources such as energy, water, and waste management (Blevis, 2007). In sustainable interaction design, researchers have increasingly put attention on studying individuals' cognitive aspects to promote behavior change towards a more sustainable way. With the understanding of psychological models, persuasive technologies (e.g., eco-feedback technologies) were developed to influence people's behaviors and to lighten their behaviors' impacts on the environment (Fogg, 2003; Froehlich et al., 2010). In addition, from a cognitive-behavioral perspective, researchers started to study the practical solutions to specific actions and objects as an alternative way in which design could promote sustainability. For example, by studying mobile phones, which are understood as disposable technology according to their "throw away and replace" culture, Huang and Truong identified opportunities and challenges for achieving the goal of sustainable design of mobile phones (Huang & Truong, 2008). This type of research is concerned with the particular relationship (potentially a sustainable relationship) between people and artifacts. The work above is focused on individuals and use whereas we aim to understand sustainable relations of groups or communities to their environments rather than unique artifacts or resources.

Everyday design

In addition to framing people as users and consumers of technologies, HCI and design researchers started to recognize the creative and sustainable ways people adapt designed artifacts (e.g., (Kim & Paulos, 2011; Roedl et al., 2015; Wakkary et al., 2013; Woodruff et al., 2008)). As a collection, these works show how repair, appropriations, and reuse through design play a contrasting role against planned obsolescence and material waste, and allow more sustainable engagements with interactive artifacts. This design strategy assumes that an empowered and skilled user can resist this obsolescence cycle through his or her individual creative acts. Wakkary and Tanenbaum (2009) have defined home dwellers as everyday designers who actively engage in the lifecycle of artifacts and use their competences towards adapting those artifacts. These practices include repair, reacquisition, and dispossession (Maestri & Wakkary, 2011; Pierce & Paulos, 2011). By supporting these practices, researchers aim to find ways to prolong the life of digital artifacts. Discussions around the empowered individual allow us to see how people have the ability to make place and to play an active role in how they construct meaning through artifacts in their environment.

Participatory and social design

Beyond focusing on the relations between individuals and artifacts, researchers also have focused on a broader view to understand people's abilities and their impacts on community contexts, which are settings in the real world encircled by place and shared experience and issues (Le Dantec, 2016). Participatory design in community settings is related to but requires different operations from the workplace, because it concerns the quality of life – better results and even more joy (Manzini, 2015) – rather than the management or labor issues present in the workplace (Ehn, 1988). To design in community settings that are explicitly pluralistic, Le Dantec proposed "Designing Publics" as a framework for social design. To address the identified issue, a public demonstrates a set of relations that are built through individuals or resources to obtain the capabilities to contend with social or political issues (Dewey, 1927). The public does not only grow out of a deliberate

A sustainable place 193

design program, but also is actively designing itself. By offering a perspective for understanding how actors, institutions, and artifacts gather around a set of issues, Designing Publics moves the objective of social design away from product development toward engaging in a process of building out infrastructure (Le Dantec, 2016). In the process of infrastructuring, publics "first articulate issues, then build out attachments, and finally integrate newly created resources for contending with issues" (Le Dantec, 2016, p. 6). It is an ongoing act of articulating and responding to issues with marshalling the available social and technical resources. Enabled by the framework of the public and its facet of infrastructuring, social design is positioned to support the development of capacities of a community to transform a set of social conditions.

Above we have presented how visions of sustainability have moved from a concern of individual behavior, to an inclusion of the empowered individual in discussions of sustainability, to a broad view of publics that contend with social issues through infrastructuring processes. While each perspective has its benefits and limitations for the design of interactive technologies, in this paper we aim at finding the right balance between perspectives of the empowered individual and the infrastructuring facet of a public as a way to discuss the concept of sustainable placemaking.

Below, we present concrete details of how this view can bring alternative understandings of sustainability by analyzing two cases: the conversion of a cargo van into a camper van and the ongoing making of a neighborhood community garden. Via two cases on the sustainable placemaking, we aim to uncover the dynamics of people's practices in making sustainable places. We believe this effort will provide a vantage point from which sustainable interaction design can operate.

Presentation of the cases

In this section, we present the two cases that we will use to investigate the idea of everyday designers as placemakers. The cases were chosen for their differences and the potential for each of the cases to reveal different insights about sustainable interaction design.

Case #1: designing, making, and living in a van interior

In the first case, we describe an autobiographical design project (Neustaedter & Sengers, 2012) of converting a cargo van into a camper van by designing and making the interior of the van (Desjardins & Wakkary, 2016). This do-it-yourself (DIY) project started in October 2013 and is still ongoing at the time of writing. Author Desjardins and her partner Bérubé LeBrun conduct the project. To date, the conversion includes the insulation of the walls, the construction of a storage platform, a bed, a table and benches, and the installation of a kitchen unit. In addition to designing and building the van interior, Desjardins and Bérubé LeBrun have experienced living in the van for over 120 days, intermittently. This autobiographical design project was previously studied to describe the qualities of living in a "prototype", or, more specifically, the qualities of designing and making a space that is lived in (Desjardins & Wakkary, 2016). We collected data through photos and time-lapse videos of the builds, Instructables[1] tutorials that were created about the build, and diary entries taken during the trips while living in the van.

This project was chosen for our analysis in this chapter for two reasons. First, it is a rare and well-documented example where a place is being designed, made, and lived in by everyday designers. In addition, this project is in the do-it-yourself (DIY) tradition and our analysis allows us to investigate deeper the relationships between DIY practices and sustainability. The DIY ethos often aims at offering an alternative to the current mode of mass production and consumption of everyday artifacts. DIY amateurs modify, hack, or make from scratch everyday artifacts in order to better fit their own needs, but also their own lifestyle and aesthetic preferences. By making an artifact specifically for themselves, DIY amateurs can not only escape the "one-size-fits-all" current model of production, but can also gain a sense of empowerment that often results in a personal connection with the artifacts they make (e.g., (Kuznetsov & Paulos, 2010; Tanenbaum et al., 2013)) or, in this case, with the place they make.

Case #2: the ongoing making of a neighborhood community garden

The second case is the ongoing making of a community garden that author Wang participated in and investigated through an ethnographic study in 2013 (Wang et al., 2015). Wang, as an international student from China, moved to Vancouver for her doctoral studies. She applied to be a member of the community garden in her neighborhood with the intention to meet more neighbors and make connections to the place that she settled in. Joining the gardening group has occurred without an explicit goal of research, similar to the autobiographical design project presented as our first case. However, community gardening participation offers a vivid perspective to look at how residents live with and shape their neighborhood environment, and furthermore, form multiple social connections. It allows us to see the detailed process of the ongoing making of a community garden and the development and maintenance of local infrastructure. In this case, we present an analysis of the making of the community gardens, inspired by the concept of placemaking. The practices of community gardeners were previously studied to understand gardeners' collaborative acts, such as information sharing, scheduling, and awareness of presence and activity (Wang et al., 2015). We collected data through observations of the gardens, participating workshops and meetings, and interviews to coordinators and members of community gardens.

In this case, we explore how gardeners develop and integrate different forms of social and technical resources. In terms of sustainability, community gardening produces an intimate relationship between people and a place (Turner, 2011), deepens people's environmental and ethical values (Hayden & Buck, 2012), and connects them with their cultural heritage (Saldivar-tanaka & Krasny, 2004). By sharing the ownership and responsibility among community members, a

A sustainable place 195

blossoming community is promoted (DeLind, 2011). We chose this case as our second case in this chapter in order to investigate how everyday designers can also participate in larger collaborative projects of placemaking in a community setting. In particular, this project highlights multiple stakeholders, diverse relationships, and the context of a broader space shared and developed by a group of neighborhood members.

Three themes of placemaking

In the next sections, we discuss three themes from our reflexive analysis of the two cases. The themes are: *longevity*, *unfinishedness*, and the *multiplicity* of everyday design strategies. These themes, as we will describe, were developed through our analysis of the cases, inspired by our readings of sustainable placemaking. We found particularly helpful the aspects of community, relationship to policy, temporality, the holistic approach, and the maintenance element in community building. We discuss how each of these themes allows a re-examination of sustainable interaction design.

Longevity

The forming of a mature and long-lasting relationship between a place and people takes time. In fact, by repeatedly re-enacting activities in a given place, people are gradually tied to the environment around them. In both cases, we observe how the use of each place is intertwined with its making and we note how longevity is as much about the making as the using or living in. The making of place arises through the long-term periods of living within that place and it is this long temporal quality that allows design acts to emerge and sustain the placemaking process.

In the van interior project, we investigate the importance of a good fit[2] for artifacts to achieve longevity in the making of a place. The van conversion project allows us to articulate and illustrate this idea, particularly with the example of the kitchen unit (Figure 11.1). The current kitchen unit is a cheap shelving unit we bought in a megastore with the intention of using it temporarily until we would design our own. We installed the shelving unit in the first months of having the van, even before finishing the walls. We had two goals when installing the kitchen unit: first to experience its presence and our use of it, and second to use it as a "sketch" of what our own handmade version would be. While we had expected to replace it rather quickly, four years later we still have it and have, in fact, started to modify and add to it (e.g., by adding a sink and pump in August 2016). To our surprise, the shelving unit lent itself very well to the creation of strong and functional groups of things in the kitchen. This inclination to support ensembles (or groups of artifacts) might explain why we kept it for much longer than we had anticipated. The metallic grid of the shelving unit allowed securing plastic bins with elastic cords, the adjustable height of the shelves allowed creation of space for the cooler, the grid above the countertop supported the drying dish towels as well as hanging lights, and the wood countertop allowed for the easy addition of a

Figure 11.1 Kitchen unit evolution

sink. The more we lived with this unit, the more we added to the "kitchen system" and the less we saw the need to change or replace the unit. In this case, the ability of the kitchen unit to truly become central to an ensemble of elements in the kitchen improved its longevity within the placemaking process.

In the community garden project, we present how a good fit also impacts the longevity of policies developed by people who act in a place. In our study on community gardens, each community garden is organized and run by an elected board of directors. Each year, they hold an annual general meeting (AGM) to summarize their work from the last year and announce the updated guidelines, policies, and responsibilities to the gardeners. We observed how gardeners adjust their rules to facilitate a better state of their collaborative work in building their garden. For instance, in one garden we studied, the coordinators found that some members were not active enough in engaging in the collaborative work. In a later AGM, the board of directors revised the collaboration rules to require gardeners to participate in work parties at least 12 hours per year to encourage greater participation. Thus, the garden members were encouraged to be involved in more communal gardening tasks than their own gardening plots. Figure 11.2 demonstrates a gardener cleaning the communal sections in the garden in a work party. The adjustment of rules was motivated by the understanding that the more the gardeners constantly and meaningfully experience the garden, the more they could understand it and the easier they could find ways to improve it.

In this case, a long period of time is needed for the gardeners to identify the changes they should make to promote a better attachment between the garden and its members. The adjustment itself also implicates their goal of prolonged community building interwoven in their everyday placemaking process.

Both the van project and the community gardens study revealed how placemaking required a long process of ongoing 'weeding' and tending to, in order to keep a healthy relationship between the place and people. While in the van project the

Figure 11.2 A gardener cleaning communal place in his garden

re-arranging and incremental changes to the place were material and physical (the kitchen unit elements), in the community gardens we saw how the re-arranging can also exist at the level of policy and human relations. In both cases, we highlight how those ongoing adjustments allow for this place to progressively function and become sustainable over time.

Unfinishedness

One of the main characteristics we have observed in our two cases is that placemaking is never finished, that elements with different levels of finish co-exist within this place. This characteristic is related to the aspect of 'longevity' presented above in the sense that the ongoing weeding and tending to assume that a place is never finished. In each case, we see how these different levels of finish co-exist and how it allows flexibility in how everyday designers live and develop a place.

We start with an example from the van conversion project. When we initiated the van project, we agreed that we would build the interior in phases. This decision was in part so that we could gather necessary resources (materials, time, and money), but also so that we could experience the place and familiarize ourselves with the evolving routines that we developed as we camped with the van at the different stages of its build.

For example, we started by building a storage platform (Figure 11.3, left), under which we could store our sport equipment and food, and on which we could eat or sleep. We used that platform as is for almost 10 months before designing

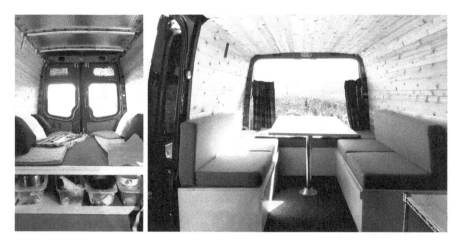

Figure 11.3 (Left) Storage and sleeping platform (Right) Benches and table built on top of the storage platform

and building a bench-table-bed unit on top of the platform to create a more comfortable eating and sleeping situation (Figure 11.3, right). The possibility to live within the place before designing every element made it possible for us to refine our idea for what that new unit should be and to make sure that it was responding to our actual needs as opposed to our idea of what a converted van should be.

In addition, in the van conversion project we observed that different elements can have different levels of finish and still co-exist in the same space. For instance, the bed-table-bench unit has reached a high level of finish, with well-varnished and polished wood together with high-quality handmade cushions. At the same time, the kitchen unit (as described in the previous section) is still only a mass-produced shelving unit we bought and installed with a few screws as a temporary 'sketch' of what our future kitchen could be. These two elements in the van can live next to each other, and their juxtaposition reminds us that this place will continue to be in flux for years to come.

Similarly, making a community garden is a process that is never finished. The stage the garden is in is always a "beta version" (Manzini, 2015, p. 52), because gardeners keep bringing new ideas and improvements to the garden, some leading to a more pleasurable experience of the garden, and others offering a more efficient and practical use of the garden. For example, members of the community garden Wang participated in recently built a fence (Figure 11.4). The original building plan of the garden did not include a fence. However, after several years, stolen produce from the garden became a constant issue. A fence with signs was built to reduce the thefts in a way that adhered to the values of the gardeners. The fence protects the gardeners' produce. Meanwhile, it keeps the garden as a place that can be experienced by all neighbors. In this way, the value of community inclusion is embodied. Therefore, the structure of the fence is uniquely designed

A sustainable place 199

Figure 11.4 Rebuilt and decorated fence in the community garden

as an ongoing project – designed to be unfinished – as values evolve. For example, children in the neighborhood decorated the fences after they were built and small plants were hung on the fence for decorations as well.

To summarize, placemaking in both cases encompasses a range of materials at different levels of age and maturity. For instance, as the storage platform became more mature in the van, it started to support new reflections for further development. In the same way, the community garden fence gained maturity as they started to be decorated and painted. In other words, 'young' solutions, like the kitchen unit in the van, are welcomed as a first step to an eventual progression and refinement over time. Overall, we see the need to reach the right balance between young materials and how their design can mature over time in order to create a sustainable place.

Multiplicity in placemaking

One of the characteristics of placemaking is to take a holistic approach to design. In our two cases, we observed how everyday designers are able to fluidly move between different strategies to make and modify their place. By adapting their strategies, they are able to address a broad range of challenges in their place. Through reflecting about their place, people actively and creatively live and experience it. Place then becomes a result of the compound function of quality and people's engagement (Casey, 2001).

We present three contrasting elements from the van interior project to show how different forms of design can take place in one project. The examples are the sketched kitchen unit (described above); the polished, thoughtful, and refined bench-table-bed unit; and the hooks added to hang lamps (Figure 11.5, left). Each of these changes in the van is a design act, however the types of planning and the required expertise in each case are dramatically different.

As a ready-made item, the kitchen unit simply required enough imagination to see the unit as a kitchen unit and to screw it to the wall. The hooks only needed to be screwed in by hand at the right position in the ceiling in order to support a wire and an LED lamp. In contrast, the bench-table-bed unit required much more thought into what was desired. Its design evolved through various paper sketches, a 3D model in SketchUp, another 3D model in Rhino (Figure 11.5, right), a

 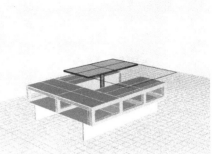

Figure 11.5 (Left) Hand-installed hooks to the ceiling (Right) 3D model of the benches and table on the platform

full-scale testing session with milk crates in the van for ergonomics, and a multi-day build to bring to life the design vision. The contrasting example of small ad hoc changes (the hooks), with appropriations of existing elements for prototyping purposes (e.g., the kitchen unit), and full 'from scratch' builds (the bench-table-bed unit) showcase the differences in the know-how necessary for each design. For instance, small ad hoc changes do not require tools; the use of only a hand suffices. Rapid prototyping and sketching in the space required ready-made elements and some connectors (screws in the case of the kitchen unit). Major changes, conversely, necessitate multiple tools, including a table saw, a miter saw, a nail gun, a drill, and varnish, and the skills to use them. In each case, we deployed the essential tools and competences for the task, but not more.

In the ongoing making of community garden projects, gardeners' strategies are diverse. For example, in addition to individual plots, gardeners also have communal plots in which they grow food together. Gardeners share harvests from the communal plot (Figure 11.6, left). Furthermore, to engage with more neighbors as well as in response to the theft problem, gardening members put part of their harvest on a shelf for sharing (Figure 11.6, right). In some community gardens in our study, members also collect coffee bean bags from local coffee shops and fill them with soil to create temporary planting beds for waiting-list residents who want to join the garden without the need to wait for a full plot. In return, the gardeners give flowers collected from the garden to the coffee shops. Similarly, community gardeners connect with local restaurants. The garden receives organic waste from the restaurants, a good fertilizer that can benefit the gardeners. In return, the restaurants can get fresh vegetables from communal plots. In this way, the community garden configures its social infrastructure. A broader network of relations was formed and served as a resource for gardeners' placemaking. In other words, the social infrastructure becomes an accessible support that enables the gardeners to act.

The two cases have shown how a multiplicity of competences, strategies, and goals co-exist as a way to support placemaking. Similar to the two previous

A sustainable place 201

Figure 11.6 (Left) Communal plot of a community garden (Right) Harvest-sharing shelf

themes, we also see how this multiplicity can address both physical skills and strategies as well as social competences and relationships. More specifically, this multiplicity of expertise is distributed across various actors who are all involved in creating a place.

Discussion

Finally, we discuss how the themes of longevity, unfinishedness, and multiplicity of design strategies offer new reflections on sustainability in interaction design. In addition, we articulate the need for a framework that explicitly describes and expresses the characteristics of placemaking for everyday designers beyond our two cases.

As a first point of discussion, we wish to highlight the important relationship between longevity and unfinishedness. The quality of unfinishedness underlines the need for a balance between young and mature materials and that balance is necessary to reach an ongoing long-lasting process of creating a place – and thus achieve longevity. In the context of sustainable interaction design, this opens new opportunities for exploring how young designs or young materials are still malleable and open to change. With refinement of the design intent, design decisions lead to the design slowly maturing, gaining higher levels of finish, and establishing a clear purpose. As our cases showed, the key is to see these young materials as elements that are ready to mature and evolve with and within the place, not as finished elements simply juxtaposed to other elements in a place.

Through the theme of multiplicity, we observed a range of strategies from tacit, to improvised and ad hoc, to fully planned. Our two cases showed how those strategies could be applied at the level of materials as well as at the level of the social and policy. What this shows is that creating and maintaining a sustainable place requires the flexibility to move between types of design actions. Specifically, this proposes that the everyday designer is able to act both in tangible and intangible realms of design, allowing for a positive synergy between artifact, environment, community, and policy. This is even more important when considering that those design actions or strategies are deployed in ongoing design processes of invariably unfinished places.

As a result of our analysis, we argue that a sustainable place is achieved through the slow and long-term design and making process, through understanding and experiencing the place in its various unfinished states, and through the flexibility and creativity of strategies – material and social – to create that place. This view of a sustainable place contributes to our understanding of sustainability by highlighting the importance of the empowered individual, but by also positioning that individual in the broader context of the place itself.

Future work

In this chapter, we have presented how the cases of the van project and the community gardens study position everyday designers as placemakers. Particularly, we have described the qualities of placemaking that are relevant to sustainable interaction design and suggested that designers and researchers can be inspired by the two cases we presented. We see this research as an important first step towards preparing a framework around sustainable placemaking for interaction design and HCI. This future framework will support investigations around how current interactive places (with various levels of computing) juggle with the ongoing making process, multifaceted everyday designers, and the relationship between artifacts and the place itself. This framework will allow us to bring more precision to a conceptual understanding of placemaking in interaction design and HCI research and projects.

Conclusion

Based on the analysis of our two cases, we articulated how a sustainable place invites people to continuously build, transform, and engage with that place and with each other in a long-term, creative, meaningful, and ongoing manner. Our hope with this chapter is to encourage interaction design researchers to rethink sustainability in terms of the creative relationship between everyday designers and a place: where the focus is broader than the individual behavior, but more constrained than reflections about the public and the social.

Notes

1 www.instructables.com/id/How-to-insulate-a-camper-van/
    www.instructables.com/id/Storage-platform-for-the-back-of-your-Sprinter-van/
    www.instructables.com/id/Cedar-paneling-for-van-interior/
    www.instructables.com/id/Bed-Table-and-Benches-for-camper-van-All-in-one/
    www.instructables.com/id/How-to-sew-cushions-for-a-camper-van/
2 The term "goodness of fit" was originally articulated by architect Christopher Alexander (Alexander, 1964) in his description of unself-conscious designers. We explore this term in more details in Wakkary, Desjardins, and Hauser (Wakkary et al., 2015).

References

Alexander, C. (1964). *Notes on the synthesis of form*. Cambridge, MA: Harvard University Press.

Blevis, E. (2007). Sustainable interaction design: Invention & disposal, renewal & reuse. In *Proceedings of the SIGCHI Conference on Human Factors in Computing Systems* (pp. 503–512). New York, NY: ACM. https://doi.org/10.1145/1240624.1240705

Casey, E. S. (2001). Between geography and philosophy: What does it mean to be in the place-world? *Annals of the Association of American Geographers*, *91*(4), 683–693.

DeLind, L. B. (2011). Are local food and the local food movement taking us where we want to go? Or are we hitching our wagons to the wrong stars? *Agriculture and Human Values*, *28*(2), 273–283. https://doi.org/10.1007/s10460-010-9263-0

Desjardins, A., & Wakkary, R. (2016). Living in a prototype: A reconfigured space. In *Proceedings of the 2016 CHI Conference on Human Factors in Computing Systems* (pp. 5274–5285). New York, NY: ACM. https://doi.org/10.1145/2858036.2858261

Dewey, J. (1927). *The public and its problems*. Denver: Swallow.

Ehn, P. (1988). *Work-oriented design of computer artifacts*. Stockholm: Lawrence Erlbaum Associates.

Fogg, B. J. (2003). *Persuasive technology*. Morgan Kaufmann. Retrieved from http://proquest.safaribooksonline.com.proxy.lib.sfu.ca/9781558606432

Franklin, A., & Marsden, T. (2015). (Dis)connected communities and sustainable place-making. *Local Environment*, *20*(8), 940–956. https://doi.org/10.1080/13549839.2013.879852

Froehlich, J., Findlater, L., & Landay, J. (2010). The Design of Eco-feedback Technology. In *Proceedings of the SIGCHI Conference on Human Factors in Computing Systems* (pp. 1999–2008). New York, NY, USA: ACM. https://doi.org/10.1145/1753326.1753629

Hayden, J., & Buck, D. (2012). Doing community supported agriculture: Tactile space, affect and effects of membership. *Geoforum*, *43*(2), 332–341. https://doi.org/10.1016/j.geoforum.2011.08.003

Heller, A., & Adams, T. (2009). Creating healthy cities through socially sustainable placemaking. *Australian Planner*, *46*(2), 18–21. https://doi.org/10.1080/07293682.2009.9995305

Huang, E. M., & Truong, K. N. (2008). Breaking the disposable technology paradigm: Opportunities for sustainable interaction design for mobile phones. In *Proceedings of the SIGCHI Conference on Human Factors in Computing Systems* (pp. 323–332). New York, NY: ACM. https://doi.org/10.1145/1357054.1357110

Kim, S., & Paulos, E. (2011). Practices in the creative reuse of e-Waste. In *Proceedings of the SIGCHI Conference on Human Factors in Computing Systems* (pp. 2395–2404). New York, NY: ACM. https://doi.org/10.1145/1978942.1979292

Kuznetsov, S., & Paulos, E. (2010). Rise of the expert amateur: DIY projects, communities, and cultures. In *Proceedings of the 6th Nordic Conference on Human-Computer Interaction: Extending Boundaries* (pp. 295–304). New York, NY: ACM. https://doi.org/10.1145/1868914.1868950

Le Dantec, C. A. (2016). *Designing publics*. Cambridge, MA: MIT Press.

Maestri, L., & Wakkary, R. (2011). Understanding repair as a creative process of everyday design. In *Proceedings of the 8th ACM Conference on Creativity and Cognition* (pp. 81–90). New York, NY: ACM. https://doi.org/10.1145/2069618.2069633

Manzini, E. (2015). *Design, when everybody designs*. Cambridge, MA: MIT Press. Retrieved from https://mitpress.mit.edu/books/design-when-everybody-designs

Neustaedter, C., & Sengers, P. (2012). Autobiographical design in HCI research: Designing and learning through use-it-yourself. In *Proceedings of the Designing Interactive Systems Conference* (pp. 514–523). New York, NY: ACM. https://doi.org/10.1145/2317956.2318034

Pierce, J., & Paulos, E. (2011). Second-hand interactions: Investigating reacquisition and dispossession practices around domestic objects. In *Proceedings of the SIGCHI Conference on Human Factors in Computing Systems* (pp. 2385–2394). New York, NY: ACM. https://doi.org/10.1145/1978942.1979291

Potter, E. (2009). A new environmental design: Sustainable place making in postcolonial Australia. *Continuum: Journal of Media & Cultural Studies, 23*(5), 697–707. https://doi.org/10.1080/10304310903180433

Project for Public Spaces. (2017). *What is placemaking?* Retrieved from www.pps.org/reference/what_is_placemaking/

Roedl, D., Bardzell, S., & Bardzell, J. (2015). Sustainable making? Balancing optimism and criticism in HCI discourse. *ACM Transactions on Computer-Human Interaction, 22*(3), 15:1–15:27. https://doi.org/10.1145/2699742

Saldivar-tanaka, L., & Krasny, M. E. (2004). Culturing community development, neighborhood open space, and civic agriculture: The case of Latino community gardens in New York City. *Agriculture and Human Values, 21*(4), 399–412. https://doi.org/10.1007/s10460-003-1248-9

Schneekloth, L. H., & Shibley, R. G. (1995). *Placemaking: The art and practice of building communities*. New York: Wiley.

Tanenbaum, J. G., Williams, A. M., Desjardins, A., & Tanenbaum, K. (2013). Democratizing technology: Pleasure, utility and expressiveness in DIY and maker practice. In *Proceedings of the SIGCHI Conference on Human Factors in Computing Systems* (pp. 2603–2612). New York, NY: ACM. https://doi.org/10.1145/2470654.2481360

Tuan, Y. (1977). *Space and place: The perspective of experience*. Minneapolis: University of Minnesota Press.

Turner, B. (2011). Embodied connections: Sustainability, food systems and community gardens. *Local Environment, 16*(6), 509–522. https://doi.org/10.1080/13549839.2011.569537

Wakkary, R., Desjardins, A., & Hauser, S. (2015). Unselfconscious interaction: A conceptual construct. *Interacting with Computers*.

Wakkary, R., Desjardins, A., Hauser, S., & Maestri, L. (2013). A sustainable design fiction: Green practices. *ACM Transactions on Computer-Human Interaction, 20*(4), 23:1–23:34. https://doi.org/10.1145/2494265

Wakkary, R., & Tanenbaum, K. (2009). A sustainable identity: The creativity of an everyday designer. In *Proceedings of the SIGCHI Conference on Human Factors in Computing Systems* (pp. 365–374). New York, NY: ACM. https://doi.org/10.1145/1518701.1518761

Wang, X., Wakkary, R., Neustaedter, C., & Desjardins, A. (2015). Information sharing, scheduling, and awareness in community gardening collaboration. In *Proceedings of the 7th International Conference on Communities and Technologies* (pp. 79–88). New York, NY: ACM. https://doi.org/10.1145/2768545.2768556

Woodruff, A., Hasbrouck, J., & Augustin, S. (2008). A bright green perspective on sustainable choices. In *Proceedings of the SIGCHI Conference on Human Factors in Computing Systems* (pp. 313–322). New York, NY: ACM. https://doi.org/10.1145/1357054.1357109

12 Interaction design for sustainability futures

Towards worldmaking interactions

Roy Bendor

A plurality of sustainability futures

Sustainability, it has been said before, may mean different things to different people.[1] Such definitional concerns notwithstanding, the position taken here is that sustainability is predominantly about the future. There is, of course, a semantic reason for that. Without the temporal elongation represented by the future, sustainability remains an empty signifier. But there are also more substantive reasons. A future devoid of nature's productive beauty is the grim image that hangs over Rachel Carson's *Silent Spring* (1962), considered by many the rallying cry of modern environmentalism,[2] just as a 'business as usual' future burdened by overpopulation, dwindling resources, and unchecked industrial growth marks humanity's endgame according to the pioneering computational models that ground *The Limits to Growth* report (Meadows et al., 1972). Inversely, prospects of a more hopeful future inspired the consolidation of sustainability as a distinct sociopolitical program. In the oft-quoted definition of sustainability (then, "sustainable development") provided by the Brundtland Commission, the future plays the role of a moral and ethical yardstick: "development which meets the needs of current generations without compromising the ability of *future* generations to meet their own needs" (UNWCED, 1987; my emphasis). Stark or hopeful, not only is the future-orientation of sustainability undeniable, but sustainability itself emerges as a platform for futurescaping – a space for imagining and materializing "hybrid, humane alternatives to the deterministic, 'business-as-usual' consensus future" (Jain et al., 2011, p. 6).

A closer look at the way the future has been mobilized in the discourse of sustainability reveals that we are actually dealing with two different, even if interrelated modalities. The first sees the future as a concrete, scientifically derived entity, a plausible historico-material outcome awaiting patiently at the present's end. Such a future is often forecasted and mobilized by policymakers. It anchors, for instance, discussions of different climate change adaptation and mitigation scenarios in reports by the United Nations' Intergovernmental Panel on Climate Change (IPCC, 2015, pp. 76–91). The second modality poses the future as an abstract, regulatory ideal – the ever-present horizon for human action. In Whitehead's (1967, p. 191) words, "Immediate existence requires the insertion of the future in the crannies of the present." In this sense, the future stands for pure possibility, representing the existential capacity of humans to act on the world intentionally, and in ways that may result in palpable change to present conditions. This future modality is more often found in philosophical, inspirational, or motivational texts such as Pope Francis' momentuous *Encyclical on Climate Change and Inequality*, where pleas for global environmental responsibility are premised in statements such as "We lack an awareness of our common origin, of our mutual belonging, and of a future to be shared with everyone" (Pope Francis & Catholic Church, 2015, p. 149). In both forms, however, the future is uncertain and may unfold in different ways; the future is plural. Accordingly, there is no single sustainable future but a plurality of sustainability futures.

This chapter asks how the plurality of sustainability futures may be communicated through the design of interactive media.[3] It then paints in broad strokes one possible answer – what, borrowing Nelson Goodman's (1978) terminology, I will call here *worldmaking interactions*.[4] These kind of interactions aim to promote the public's own ability to imagine alternative futures – to encourage the public to find ways to collectively reformulate a sense of what is possible and hopefully rediscover its capacity to critically make, unmake, and remake the world. But before I discuss worldmaking interactions, a few more general remarks about interaction design for sustainability are warranted.

Design as futurescaping

Like sustainability, design can also be seen as a form of futurescaping.[5] But in what sense? In a rather straightforward manner, to design an artifact, service, or interaction is to foreshadow a future where the artifact, service, or interaction will be used to accomplish something or be meaningful for someone. Thus, as Victor Papanek (1972, p. 3) famously remarked, the design process includes "The planning and patterning of any act towards a desired, foreseeable end". We see this future-orientation in practical design frameworks such as the Vision in Product Design (VIP) approach, where considerations of future contexts, interactions, and outcomes occupy the latter half of the model design process (Hekkert & Van Dijk, 2011). But I want to suggest here that design is a form of futurescaping not only teleologically – by prescribing future uses, situations, or relations – but also by conjuring images of possible futures, that is, by occupying the very space that lies between the two notions of the future noted above. Design, in other words, concretizes ideal images of the future into material future-instances, while deriving from the latter new meanings, significances, and values that inform the ideal future. Blevis (2007, p. 503) hints at this when he defines design as "an act of choosing among or informing choices of future ways of being" (emphasis removed), and so does Marenko (2016, p. 2758) when she argues that design is "the practice of materialising possibilities". On both accounts, design can be seen as a means for proliferating futures.

Given that both sustainability and design share a deep future-orientation, in practices that overlap the two (such as interaction design for sustainability) the future is constantly present. Even if this is not always made explicit, the future appears under two different aspects: first as part of the designer's toolbox, and second as a subject area for design interventions.

As part of the designer's toolbox, the future augments processes of interaction design for sustainability with considerations of long-term impacts, alternative use contexts, or otherwise helps to expand the "technological design horizon" (Blyth et al., 2015). Thus, Mankoff et al. (2013, p. 1631) write that "Futures Studies methods can help to question what to design, as well as what aspects of contemporary designs are likely to resonate with (or against) future trends." Tanenbaum et al. (2013, p. 33) find inspiration in the Steampunk subculture for "a future in which design is driven by aesthetics, grounded in a sustainable ethos, and aimed at serving the needs and preferences of distributed communities of engaged expert users". In a more recent article, Tanenbaum et al. (2017) argue that design fiction, "Positioning an imagined technology within a narrative world" (p. 64), presents interaction design for sustainability with an opportunity to reflect on the kind of values, arguments, and forms of engagement it produces, especially in relation to popular culture. Reflection is also at the core of Wakkary et al.'s (2013, p. 2) suggestion that design fiction "can be readily incorporated into [design] practices in ways that transform those practices", providing designers with cues for "material reflection that is based in making and doing, where scenarios, prototypes, sketches, and illustrations are materials of thought for design". Whether it is part of design fiction, speculative design, or material speculation, the future may inspire designers to expand their purview both outwards and inwards.

Under the second aspect, the future unfolds as a thematic space for a range of specific designs for sustainability. Such designs may seek to influence unsustainable behavior, achieve learning goals related to sustainability, or evoke a sense or experience of sustainability. Bonnie Nardi (2016, p. 27) captures the sentiment of many such activities when she points out that since "Technology drives change ... we in human-computer interaction should attend to the future." Her subsequent question, "But which future? The one we want? The one we think is coming?" (ibid.), has been mostly answered by the sustainable HCI community with a dose of skeptic realism. Tomlinson et al. (2012), for instance, propose "collapse informatics" as a response to a future characterized by the unraveling or decomplexification of society. It may not be the future they want, but it certainly is the one they think we should prepare for. Similarly, efforts to rethink "computing within limits" that developed from the collapse informatics approach use scientific accounts of the planet's diminishing resources as a hard, "non-negotiable" container for sociotechnical solutions. A future projected from a sober "ecological reality", as Pargman and Raghavan (2014, p. 645) hold, is posited as the alpha and omega of sustainable HCI, making way for a view of sustainability as "revolutionary" not by its capacity to reimagine the world but by its ability to adapt "to a reality of limits, of trade-offs, and of hard choices".

There is certainly good reason to respond to Nardi's (2016) question ("But which future?") with a measure of realism, after all, despite some success, prospects for keeping climate change below the key threshold of 2°C above pre-industrial temperatures seem less and less likely in our current political climate. However, I fear that clinging to a scientifically-derived notion of what is and what is not possible may actually limit the range of possible futures for which we may design, and effectively, even if inadvertently, promote what futurist Stuart Candy (2008) calls "monofuturism": the "fundamental yet distressingly widespread misconception that engaging the yet-to-be means trying to predict 'the future' rather than exploring alternative futures". This isn't to say that there is no value to scientific accounts of environmental dynamics and trajectories, that we should jettison scientific epistemology or operate under unrealistic assumptions that no ecological limits exist. But if interaction design for sustainability is to succeed as a platform for futurescaping, shouldn't designers work to breach the confinements of *any single future* – whether it be utopian or dystopian, hopeful or desperate, cornucopian or marred by scarcity? In this sense, Nardi's question may be pluralized to ask, "Which futures?", and the answer proposed here would be, "As many as possible!" What I am suggesting, then, is that the task of interaction design for sustainability could be seen as similar to what the protagonist in Ursula Le Guin's novel, *The Dispossessed*, identifies as the job of the "thinking man": "not to deny one reality at the expense of the other, but to include and to connect".[6]

Scientifically defensible and diegetically consistent futures

Several forms (or genres) of interactive media for sustainability already attempt to conjure, concretize, and multiply different images of the future. Perhaps the most well known of these are simulation games such as SimCity. Through SimCity's isomorphic landscape and informative dashboard, urban sustainability appears as a complex, emergent system. Users are invited to understand the system by making a series of choices among diverse urban planning variables (density of built environment, methods of energy production, use of resources, etc.) and then witness the impact of their choices on the entire system over time. What simulation games such as SimCity do, in other words, is provide users with the means to multiply and compare futures based on different inputs. Repeating this process is said to promote a deeper understanding of the various dimensions of sustainability, the trade-offs involved in complex sustainability decision-making, the challenges in balancing short- and long-term imperatives, and the ability to generalize from particular instances to the system as a whole (Gaber, 2007; Kim & Shin, 2016; Robinson et al., 2011; Rothman et al., 2002; Terzano & Morckel, 2016). Significant for our purposes here, however, is the way sustainability simulations tend to pursue scientific "defensibility" by trying to maintain fidelity to current scientific understandings of social and natural phenomena (Sheppard, 2001). As told by SimCity's creative director, Ocean Quigley: "We're doing our best to model real systems . . . so that you'll understand something of how they actually work. And you'll make the tradeoffs that real cities have to make."[7] Futures generated by sustainability simulations, therefore, rely on scientific data to manufacture and retain believability, credibility, and trust.

Where sustainability simulations seek consistency with science, other, more immersive media seek a different form of consistency, one in which believability is not produced as a function of model-reality relations but as a quality of the relations between the different elements within the immersive environment. They aim, in other words, for diegetic consistency – the sense in which elements within a fictional world seem to fit together. Take for example the 'guerrilla art' installation that was commissioned by Canadian environmental group, the Dogwood Initiative, in 2014. The installation, part of the group's #NoTankers campaign, invited the public to peer into responsive binoculars located in Vancouver's English Bay and witness a 3D virtual reality depiction of the effects of a disastrous oil spill on the beach, including a sinking tanker, billowing smoke, tar covering the shoreline, and a beached Orca whale. Any effect produced by the immersive environment glimpsed through the binoculars was not necessarily a product of the installation's scientific defensibility. It didn't really matter whether an oil spill in the bay would indeed cause all the effects suggested by the installation, or would cause those effects in the exact manner in which they were depicted. What mattered was the degree to which the different elements fit together into a consistent world, conveyed a believable story, and brought to life a provocative possible future.[8] Drawing from Candy (2010), we can say that when such immersive media are successful, they transform sustainability futures into a space to inhabit, a situation to negotiate, unpack, or work through. So while sustainability simulations conjure multiple futures based on a numeric or statistical forms (as models, data points, and so forth), immersive media conjure multiple futures as a space of semiosis and diegesis. Simply stated, they aim to provoke a felt relation to the future. Dogwood Initiative's Energy and Democracy Director at the time, Kai Nagata, admits as much: "We wanted to bring it home, and give people a little flavour of what this could look like here in our city."[9]

While both sustainability simulations and sustainability-related immersive environments multiply futures, neither allows users to imagine the future on their own terms. Instead, users are invited to step into futures already imagined for them by designers. There are, of course, practical advantages to this, especially in terms of scaling up the interactive intervention. But there are also ethical questions that emerge from the perceived authority of computational systems. If the scientific premise of sustainability simulations confers a sense of truth to the futures they unfold, and if the diegetic consistency of immersive sustainability futures results in a strong sense of believability, wouldn't such media end up reinforcing "monofuturism" despite their best intentions? This possibility itself, I believe, is a sufficient incentive to explore and experiment with new forms of interactions for sustainability.

Worldmaking interactions

Worldmaking interactions emerge from the proposition that our approach to sustainability is always already premised in the way we understand the world, the entities that inhabit it, and the relations that bind them (Maggs & Robinson, 2016). Sustainability, in other words, is understood ontologically as a matter of deeply held beliefs about the world, inseparable from the most elemental ways by which we perceive the world and render it meaningful. Accordingly, worldmaking interactions seek to make visible and therefore malleable the connections between those deeper ways of relating to the world and a multiplicity of sustainability futures. They strive, in Tony Fry's (2009, p. 11) words, to potentiate "an ontological shift in the mode of being of the actor". They are neither prescriptive nor didactic; they seek not to facilitate the user's capacity to 'solve' sustainability futures, nor to evoke a sustainability-related "future shock". Instead, they target a deeper layer of our relation to sustainability futures.

One of the elements that make worldmaking interactions interesting is the emphasis they put not on encountering sustainability through scientific facts and 'appropriate' technological fixes – markers of the standard "epistemological stance of sustainability" according to Alaimo (2012, p. 561) – but through the co-articulation of social and cultural concerns. Instead of focusing on the environmental impacts of sustainability (or, more often, the material consequences of unsustainability), they suggest that since the world is a social creation we can all take part in its making, unmaking, and remaking (Goodman, 1978; Watzlawick, 1977). *We are all worldmakers*. Given that this process involves not only "objective facticity" but also "subjective meaning" (Berger & Luckmann, 1989), worldmaking interactions draw from artistic vocabularies and practices that make use of playfulness, ambiguity, and open-endedness as communicative vehicles. Instead of placing users within pre-existing sustainability futures, be they based on scientific or speculative representations of the world, worldmaking interactions only hint at those futures, leaving them unarticulated or 'unfinished'. In this sense, worldmaking interactions draw from both research that points to the important role the imagination plays in engaging with sustainability (Wright et al., 2013; Yusoff & Gabrys, 2011), and from recent calls to consider ambiguity, uncertainty, and discomfort not as hindrances but as assets to impactful interactive design (Benford et al., 2013; Cox et al., 2016; Dunne & Raby, 2013; Fuad-Luke, 2009; Marttila, 2011).

Along with a multidisciplinary team of artists, designers, and sustainability practitioners, I was recently involved in a pilot attempt to materialize worldmaking interactions by creating a multimedia installation called *Sustainability in an Imaginary World*.[10] The installation combined a narrative, four interactive spaces, and several choice-making activities. It involved digital and analog interactivity, a full theatrical set, and a live actor. After watching a short fake news segment that evokes the more standard narrative of a future characterized by extreme climatic events, social unrest and political collapse, participants (in groups of six) were ushered by an actor dressed as a janitor into the installation's main room. The dark, sparing industrial space included a few wooden crates. Participants could then hear three different voices, each describing the same climatic events and suggesting a solution from their own perspective. Soon after, one of the crates reveals itself as a touch-screen table, inviting participants to make two choices about how to best address the crisis: Should we rely on technological innovation or on changes in human behavior and lifestyle? And should we seek collective solutions or prompt individual action? Aggregated choices dissolve into an image that symbolically represents the nature of the group's choice. The same image will accompany the group as they explore three different rooms, whose presence is now disclosed only by a dim light that creates door shapes on the tarps that surround the main space. Participants are not given additional cues or instructions; they may or may not enter the three rooms, may do so together or alone, and in the sequence of their choice.

The three rooms that open up to the main space correspond to three worldviews: a spiritual room, a materialist room, and a literary room.[11] Each room features light effects and projections; ambient sound and the audial presence of one of the three characters encountered in the main space; a representation of a tree; digital and analog interactive elements. In the white-colored spiritual room, participants encounter a circular path that surrounds a blanched tree whose branches meet another blanched tree that hangs from the ceiling. Masks hanging on the wall invite participants to engage in an unspecified ritual. The copper-colored materialist room features a Steampunk aesthetic. At its center stands a metallic tree decorated with locks, keys, chains, and pulleys. When the locks are opened and the chains pulled, the tree is made to blossom. In the literary room a multitude of doors open to a curiosity cabinet, regular and funhouse mirrors, live closed-circuit camera feeds, and a tree sectioned inside stacked jars. Green leaves are strewn on the floor.

After participants have the opportunity to wander in and between the rooms, the janitor returns to nudge them back to the lobby. But before he manages to do so, a hanging crate descends from the ceiling and reveals dozens of white, copper, and green leaves – color-matched to the three rooms. A voice invites participants to make a choice about the best way to address the situation: "White leaves for the knowledge that something is out there. Copper leaves for the faith that answers will come. Green leaves for the comfort in knowing no world is carved in stone." With every leaf plucked, a representation of the same-colored tree is projected on the walls of the room, resulting in a forest of trees that manifests the group's collective choice.

As mentioned above, playful ambiguity and open-endedness are key ingredients of worldmaking interactions. They are present in the installation in two important ways. First, the installation attempts to create productive ambiguity by refusing to give participants clear instructions about how to move through the space, or how to interact with the various elements. However, although participants receive very little guidance, common threads are woven throughout the interactive experience to provide users with the relative safety of a banister as they process the more 'risky' aspects of the experience. In order to create continuities between the obvious and the imaginary, the familiar and the strange, the installation relied on a narrative, complete with consistent colors, objects, symbols, and voices, to link the otherwise rather open-ended spaces of which it was comprised. In this mode, we aimed to leave several 'blank canvases' on which users could project their own imagination, while providing them with enough of a through line to be able to follow the experience to its conclusion.

Second, the installation made playful use of the differences between the busier intentionality that often emerges with interactivity and the reflexivity necessary for deep contemplation. Participant disposition, in this sense, became another of the installation's building blocks, material for designers to work with. Of course, participant dispositions are colored by previous experiences with interactive systems, but also by pre-existing conceptions of sustainability. As we found out in post-experience interviews, many participants arrived at the installation armed with an understanding of sustainability as a complex problem to solve, and were therefore expecting to encounter a series of interactive tasks that would be equivalent to achieving or mastering sustainability futures. The installation pre-emptively addressed such expectations by combining the more freeform wandering that took place in and between the installation's abstract spaces with moments of a more task-oriented interactivity that included achievable goals. A similar effect was sought by the use of narrative. After the installation established a fairly conventional narrative – an account of a dystopian future underlined by climatic emergencies and political upheaval – it shifted registers and drew participants into a more contemplative, abstract, and unguided experience. In this way we tried to harness the experiential value of surprising users and providing them with more "complicated pleasures" (Dunne & Raby, 2001). This was justifiable, to some extent, by accounts that suggest that moments of confusion, disappointment, and even boredom can be important triggers for deep reflection and creativity (Svendsen, 2005; Toohey, 2011).

Since *Sustainability in an Imaginary World* is only an early attempt at conceptualizing and applying worldmaking interactions, we should be careful not to overstate the installation's outcomes (positive or negative). Not all those who visited the installation came out of the experience with a deeper or even different understanding of sustainability, or with a clear idea about what they have experienced at all. Despite our best intentions, some visitors were utterly confused, frustrated, and even agitated (which led us to add group discussions with participants after they exited the installation). Yet others reported feeling emotional and reflexive, were ready to examine the way they have previously considered sustainability, and were willing to engage positively with the kind of ontological multiplicity the installation suggested. They felt, in other words, that worldmaking was indeed within their power.

"The future cannot be predicted, but futures can be invented"[12]

The point of departure for this chapter is that both sustainability and design represent humanity's capacity to shape the future in meaningful ways, that they are forms of futurescaping. Interaction design for sustainability can play a significant role in this process by infusing scientifically conceived projections of the future with cultural sensitivities, imagery, symbolism, and meaning. As more people take part in interactive futurescaping, the process of futurescaping itself may become more inclusive, helping to democratize the future, as Candy (2010, p. 16) puts it. There is certainly support for such an agenda outside of the design field, where some have already pointed at the "staggering lack of imagination in our own time" (Williams & Srnicek, 2013) as part of our inability to respond appropriately and swiftly to the great challenges of our time (Augé, 2015; Ghosh, 2016; Haiven, 2014; Klein, 2014; Yusof & Gabrys, 2011). Given that one of the keys to the success of past transformative social movements has been the way they "dreamed in public, showed humanity a better version of itself, modeled different values in their own behavior, and in the process liberated the political imagination and rapidly altered the sense of what was possible" (Klein, 2014, p. 462), transformative change depends on our capacity to expand our social and political imaginaries – a capacity whose atrophy seems to herald the Anthropocene.[13]

Our chances of surviving the Anthropocene will surely improve if the future itself could be reclaimed and the nefarious dominance of short-termism in modern life – the vicious cycle of "defuturing" as Fry (2009) puts it – broken. Interaction design for sustainability can support this by proliferating diverse, thought-provoking, inspiring futures, but also by helping to develop the capacity of the public to multiply, discuss, and evaluate futures on their own. This way we may, perhaps, "borrow the energy from the future to overturn the conditions of the present", as Braidotti asks.[14] I have suggested here that worldmaking interactions may illustrate a path for such a project by provoking users to engage their own imagination as a way to pluralize and problematize any single view of sustainability and any singular version of a sustainable future. The world they manifest is malleable and invites us all to shape and reshape it together. In this sense, worldmaking interactions propose future-thinking and futurescaping as fundamental, emancipatory social activities.

Notes

1 For a few examples see Alaimo, 2012; Connolly, 2007; Ehrenfeld, 2008; Jacobs, 2006; Miller, 2013.
2 See for instance Dauvergne, 2009, p. 26.
3 To be clear, I refer to interactivity here not as a technical property but as an experiential quality, the corollary of a "responsive aesthetic" (Krueger, 2003).
4 For a more detailed explanation of Goodman's thesis and a view of worldmaking in the context of scenario analysis see Vervoort et al., 2015.
5 See for instance Fry, 2009; Goodman, 2008; and the essays collected in Yelavich and Adams, 2014. The title of the 50th anniversary conference of the Design Research Society (DRS'16), "future-focused thinking", is another indication of the fundamental role the future plays in design research and practice, as is the recent special section dedicated to "everyday futures" in the influential design journal, *Interactions* (March-April, 2017).
6 Le Guin, 1974, p. 251.
7 Quigley is quoted in Massey, 2012.
8 Adrian Crook, one of the installation's creators, described it as a "location based 'stunt' to generate publicity" (Crook, 2014).
9 Nagata is quoted in Ball, 2014.
10 The installation was funded by a three-year Insight grant from the Canadian Social Sciences and Humanities Research Council (SSHRC). The discussion of the installation

presented here is drawn from an earlier account of the installation's design (Bendor et al., 2015) and a more recent analysis of its pilot run (Bendor et al., 2017).
11 The choice of worldviews was inspired by Rorty, 2007.
12 Gabor, 1964, p. 161.
13 On social imaginaries see Taylor, 2004.
14 See Braidotti & Vermeulen, 2014.

Resources Used

Alaimo, S. (2012). Sustainable this, sustainable that: New materialisms, posthumanism, and unknown futures. *PMLA, 3*(127), 558–564.

Augé, M. (2015). *The future*. London, New York: Verso.

Ball, D. P. (2014, November 12). *Burrard inlet binoculars imagine oil-slicked disaster*. Retrieved April 22, 2017, from http://thetyee.ca/News/2014/11/12/Burrard-Inlet-Installation

Bendor, R., Anacleto, J., Facey, D., Fels, S., Herron, T., Maggs, D., . . . Williams, S. (2015). Sustainability in an imaginary world. *Interactions, 22*(5), 54–57.

Bendor, R., Maggs, D., Peake, R., Robinson, J., & Williams, S. (2017). The imaginary worlds of sustainability: Observations from an interactive art installation. *Ecology and Society, 22*(2), article #17.

Benford, S., Greenhalgh, C., Giannachi, G., Walker, B., Marshall, J., & Rodden, T. (2013). Uncomfortable user experience. *Communications of the ACM, 56*(9), 68–73.

Berger, P. L., & Luckmann, T. (1989 [1966]). *The social construction of reality: a treatise in the sociology of knowledge*. New York: Anchor Books.

Blevis, E. (2007). Sustainable interaction design: Invention & disposal, renewal & reuse. In M. B. Rosson & D. Gilmore (Eds.), *Proceedings of CHI 2007* (pp. 503–512). New York: ACM Press.

Blyth, P., Mladenović, M., Nardi, B., Su, N., & Ekbia, H. (2015). Driving the self-driving vehicle: Expanding the technological design horizon. *Proceedings of the IEEE International Symposium on Technology in Society (ISTAS)*.

Braidotti, R., & Vermeulen, T. (2014, August 12). *Borrowed energy*. Retrieved April 22, 2017, from www.frieze.com/issue/article/borrowed-energy

Candy, S. (2008). *Object-oriented futuring*. Retrieved April 23, 2017, from https://futuryst.blogspot.nl/2008/11/object-oriented-futuring.html

Candy, S. (2010). *The futures of everyday life: Politics and the design of experiential scenarios*. PhD Dissertation. University of Hawaii, Manoa.

Carson, R. (1962). *Silent spring*. Boston, MA: Houghton Mifflin; Riverside Press.

Connolly, S. (2007). Mapping sustainable development as a contested concept. *Local Environment: The International Journal of Justice and Sustainability, 12*(3), 259–278.

Cox, A. L., Cecchinato, M. E., Gould, S., Iacovides, I., & Renfree, I. (2016). Design frictions for mindful interactions: The case for microboundaries. *Proceedings of CHI'16 (Extended Abstracts)*, May 07–12, San Jose, CA.

Crook, A. (2014, November 20). *How AC+A created a (virtual reality) oil spill in just 6 weeks*. Retrieved April 22, 2017, from http://adriancrook.com/how-we-delivered-a-virtual-reality-installation-in-just-6-weeks/

Dauvergne, P. (2009). *Historical dictionary of environmentalism*. Lanham, MD: Scarecrow Press.

Dunne, A., & Raby, F. (2001). *Design Noir: The secret life of electronic objects*. Basel: August/Birkhäuser.

Dunne, A., & Raby, F. (2013). *Speculative everything: Design, fiction, and social dreaming*. Cambridge, MA: MIT Press.
Ehrenfeld, J. (2008). *Sustainability by design: A subversive strategy for transforming our consumer culture*. New Haven: Yale University Press.
Fry, T. (2009). *Design futuring: Sustainability, ethics, and new practice*. Oxford; New York: Berg.
Fuad-Luke, A. (2009). *Design activism: Beautiful strangeness for a sustainable world*. London; Sterling, VA: Earthscan.
Gaber, J. (2007). Simulating planning – SimCity as a pedagogical tool. *Journal of Planning Education and Research*, *27*(2), 113–121.
Gabor, D. (1964). *Inventing the future*. New York: Alfred A. Knopf.
Ghosh, A. (2016). *The great derangement: Climate change and the unthinkable*. Chicago IL & London: The University of Chicago Press.
Goodman, D. (2008). *A history of the future*. New York: Monacelli Press.
Goodman, N. (1978). *Ways of worldmaking*. Indianapolis: Hackett Pub. Co.
Haiven, M. (2014). *Crises of imagination, crises of power*. London: Zed Books.
Hekkert, P., & Van Dijk, M. B. (2011). *Vision in design: A guidebook for innovators*. Amsterdam: BIS.
IPCC. (2015). *Climate change 2014: Synthesis report. Contribution of working groups I, II and III to the fifth assessment report of the intergovernmental panel on climate change*. IPCC: Geneva, Switzerland. Retrieved April 22, 2017, from: www.ipcc.ch/pdf/assessment-report/ar5/syr/SYR_AR5_FINAL_full_wcover.pdf
Jacobs, M. (2006). Sustainable development as a contested concept. In A. Dobson (Ed.), *Fairness and futurity: Essays on environmental sustainability and social justice* (pp. 21–45). Oxford: Oxford University Press.
Jain, A., Ardern, J., & Pickard, J. (2011). Design futurescaping. In *Blowup Reader#3: The era of objects* (pp. 6–14). Rotterdam: V2.
Kim, M., & Shin, J. (2016). The pedagogical benefits of SimCity in urban geography education. *Journal of Geography*, *115*(2), 39–50.
Klein, N. (2014). *This changes everything: Capitalism vs. the climate*. Toronto: Alfred A. Knopf.
Krueger, M. W. (2003). Responsive environments. In N. Wardrip-Fruin & N. Montfort (Eds.), *The new media reader* (pp. 379–389). Cambridge, MA & London: MIT Press.
Le Guin, U. K. (1974). *The dispossessed; an ambiguous Utopia*. New York: Harper & Row.
Maggs, D., & Robinson, J. (2016). Recalibrating the anthropocene: Sustainability in an imaginary world. *Environmental Philosophy*, *13*(2), 175–194.
Mankoff, J., Rode, J. A., & Faste, H. (2013). Looking past yesterday's tomorrow: Using futures studies methods to extend the research horizon. *Proceedings of CHI 2013, April 27–May 2, 2013*, Paris, France, 1629–1638.
Marenko, B. (2016). Introduction: Design-ing and creative philosophies. In P. Lloyd & E. Bohemia (Eds.), *Proceedings of DRS16: Design + Research + Society – Future-Focused Thinking* (Vol. 7, pp. 2757–2760).
Marttila, T. (2011). Unpleasurable products and interfaces: Provocative design communication for sustainable society. *Proceedings of DPPI '11*, June 22–25, Milano.
Massey, N. (2012, March 12). *SimCity 2013 players will face tough choices on energy and environment*. Retrieved April 22, 2017, from www.scientificamerican.com/article/simcity-2013-players-face-tough-energy-environment-choices
Meadows, D. H., Meadows, D. L., Randers, J., & Behrens III, W. W. (1972). *The limits to growth; a report for the Club of Rome's project on the predicament of mankind*. New York: Universe Books.

Miller, T. R. (2013). Constructing sustainability science: Emerging perspectives and research trajectories. *Sustainability Science, 8*(2), 279–293.
Nardi, B. (2016). Designing for the future – but which one? *Interactions, 23*(1), 26–33.
Papanek, V. J. (1972). *Design for the real world: Human ecology and social change*. New York: Pantheon Books.
Pargman, D., & Raghavan, B. (2014). Rethinking sustainability in computing: From Buzzword to bon-negotiable limits. *Proceedings of NordiCHI '14*, October 26–30, Helsinki, Finland, 638–647.
Pope Francis & The Catholic Church. (2015). *Encyclical on climate change and inequality: On care for our common home*. The Vatican, Rome.
Robinson, J., Burch, S., Talwar, S., O'Shea, M., & Walsh, M. (2011). Envisioning sustainability: Recent progress in the use of participatory backcasting approaches for sustainability research. *Technological Forecasting and Social Change, 78*(5), 756–768.
Rorty, R. (2007). Philosophy as a transitional genre. In *Philosophy as cultural politics* (pp. 3–28). Cambridge: Cambridge University Press.
Rothman, D. S., Robinson, J. B., & Biggs, D. (2002). Signs of life: Linking indicators and models in the context of QUEST. In H. Abaza & A. Baranzini (Eds.), *Implementing sustainable development, integrated assessment and participatory decision-making processes* (pp. 182–199). Cheltenham, UK: Edward Elgar.
Sheppard, S. R. J. (2001). Guidance for crystal ball gazers: Developing a code of ethics for landscape visualization. *Landscape and Urban Planning, 54*(1), 183–199.
Svendsen, L. F. H. (2005). *A philosophy of boredom*. London: Reaktion Books.
Tanenbaum, J., Desjardins, A., & Tanenbaum, K. (2013). Steampunking interaction design: Principles for envisioning through imaginative practice. *Interactions, 20*(3), 29–33.
Tanenbaum, J., Pufal, M., & Tanenbaum, K. (2017). Furious futures and apocalyptic design fictions: Popular narratives of sustainability. *Interactions, 24*(1), 64–67.
Taylor, C. (2004). *Modern social imaginaries*. Durham, NC: Duke University Press.
Terzano, K., & Morckel, V. (2016, March 7). SimCity in the community planning classroom: Effects on student knowledge, interests, and perceptions of the discipline of planning. *Journal of Planning Education and Research*.
Tomlinson, B., Silberman, M. S., Patterson, D., Pan, Y., & Blevis, E. (2012). Collapse informatics: Augmenting the sustainability & ICT4D discourse in HCI. *Proceedings of CHI'12*, May 5–10, Austin, TX, 655–664.
Toohey, P. (2011). *Boredom: A lively history*. New Haven: Yale University Press.
UNWCED. (1987). *Our common future: Report of the world commission on environment and development*. New York City: Oxford University Press.
Vervoort, J. M., Bendor, R., Kelliher, A., Strik, O., & Helfgott, A. (2015). Scenarios and the art of worldmaking. *Futures, 74*, 62–70.
Wakkary, R., Desjardins, A., Hauser, S., & Maestri, L. (2013). A sustainable design fiction: Green practices. *ACM Transactions on Computer-Human Interaction, 20*, 4. Article No. 23.
Watzlawick, P. (1977). *How real is real? Confusion, disinformation, communication*. New York: Vintage Books.
Whitehead, A. N. (1967). *Adventures of ideas*. New York: The Free Press.
Williams, A., & Srnicek, N. (2013). # Accelerate manifesto for an accelerationist politics. *Critical Legal Thinking, 14*, 72–98.
Wright, C., Nyberg, D., Cock, C. D., & Whiteman, G. (2013). Future imaginings: Organizing in response to climate change. *Organization, 20*(5), 647–658.
Yelavich, S., & Adams, B. (Eds.). (2014). *Design as future-making*. London & New York: Bloomsbury.
Yusoff, K., & Gabrys, J. (2011). Climate change and the imagination. *Wiley Interdisciplinary Review: Climate Change, 2*, 516–534.

13 Think local act local

The case of Burning Man

a.m. tsaasan and Bonnie Nardi

In June 1986, two artists from the San Francisco Bay Area built "The Man", an eight-foot wooden human effigy that was set alight in a bonfire ring at Baker Beach on the northern coast of California. This "burn" drew a group from the surrounding area to watch the fire. For the next few years, increasing numbers of people returned to Baker Beach each June with various forms of flammable art to add to the evolving burning of that year's incarnation of The Man. The participatory event came to be called Burning Man.

The community emerging from the Baker Beach burns embodied sustainability from the outset by upcycling landfill diversion materials into art before burning. Some pieces at Burning Man were created for engagement with the public during the day and then burned, and others were created solely for the pleasure of watching them burn. Artists often designed their work to guide flames along materials to achieve particular shapes or negative spaces. In the early years, all art installations at the Burning Man event were created with the intent to burn. Today, very few large-scale installations are designed to burn, and most are reused in some form. For more than 20 years, the first author has created and co-created interactive installations at Burning Man community events both for burning and for reuse.

The image on the left in Figure 13.1 depicts the 2015 mini-effigy burn, traditionally held at sunset on the same night as the main effigy burn for a Southern California community's annual Burning Man Regional Event. The image on the right is the mini-effigy creation in process. Here the first author's son sits amid the partially constructed mini-effigy and various post-construction salvaged materials, including those from the main effigy creation process. This installation was primarily directed by eight children between 5 and 13 years of age, with adults, including the first author, in supportive roles.

This chapter examines technological contexts supporting the growth of local, geographically bound, culturally appropriate community sustainability practices derived from the ethos informing the annual Burning Man event. We examine the San Francisco–based non-profit organization Burning Man Project's outreach program known as the Burning Man Regional Network, and the ways in which local collaboration, action, and innovation *are supported through human networks leveraging older digital technologies.* Four communities are located in Nevada and three in California.

Figure 13.1 Mini-effigy burn and creation
Photo credit: Heather L. Hicks, 2015

Burning Man takes place annually in September in Nevada (having moved from California in 1990), and invites interactive public art in various forms. Though fewer pieces are set alight, it is still one of the largest venues on earth supporting burning art, which remains a hotly debated ephemeral art form. Public bonfire practices have been documented in early human communities. Bonfire traditions persist in communities around the world. For example, today bonfires are built for the purpose of community interaction in Persian communities around the world in honor of the annual Chaharshanbe Suri, which "dates back to the year 1725 B.C. and is believed to be an annual ceremony of ancient Persians. At this festival, ancient Persians would set fires and jump over them..." (Tavakoli et al., 2011). The image in Figure 13.2 was taken in 1991 by multimedia artist William Binzen. It depicts a crowd gathered for the climatic interactive art performance of the week-long Burning Man event.

Over the past 30 years, tens of thousands of burners have participated in "a temporary metropolis dedicated to community, art, self-expression, and self-reliance [in which t]hey depart one week later, having left no trace whatsoever" (Burning Man Project, 2017a). While participants "leave no trace" in the Nevada desert, many bring the leave no trace ethos to their local communities in the form of culturally appropriate changes to local practices. For example, the term "MOOP" refers to "Matter Out Of Place" and directs attention to the appropriateness of an object in relation to its geographical location. The ethos suggests personal attention, responsibility, and action to remove any object that is *out of place*.

The practices examined here, while inspired by necessity at Burning Man and supported indirectly by the Burning Man Project, are uniquely suited to the local communities in which they are enacted. These practices are *local* in that they are sensitive to and interact with the local communities, resources, tools, and

Think local act local 219

Figure 13.2 A crowd gathered for the Burning Man
Photo credit: William Binzen, 1991

contexts. They are local in that they *do not seek to replicate or generalize practices to other environments*. Instead, there appears to be a transfer of the Burning Man *ethos* emphasizing communal effort and personal and civic responsibility into local communities.

Local practices are the antithesis of "best practices", "portable practices", and other methodologies for moving toward the standardization of a one-size-fits-all world. The Burning Man Project "History of Regionals" online narrative timeline emphasizes that "[E]ach and every [region] possesses a unique flavor and character." From brick and mortar build spaces to participation in local civic events to the creation of local not-for-profit service organizations, the Burning Man Regional Network supports any local cultural form of expression that does not infringe on trademarked or copyrighted material and that embodies the Burning Man ethos. Formalized in 2004, the ethos comprises Ten Principles: radical inclusion, giving, decommodification, self-reliance, self-expression, communal effort, civic responsibility, leaving no trace, participation, and immediacy. The interpretation and manifestation of these principles varies widely, even within burn communities near each other.

Throughout this chapter we will be using community terminology including "The Man", a human effigy redesigned each year to burn at the annual week-long Burning Man event in the location "Black Rock City", which is geographically located in the federally managed Black Rock Desert-High Rock Canyon Emigrant Trails National Conservation Area in northern Nevada. Black Rock City is commonly referred to as "home" in contrast to the "default" world in which most participants live the other 51 weeks a year. Practicing and/or demonstrating fluency in the Ten Principles is sufficient for a self-identified "burner" to be considered a "participant" in a "burn community." Inclusion does not necessitate ever having attended the event in Black Rock City, and some very active burners in the seven burn communities included in this chapter have never attended the "main" burn, a sanctioned "Burning Man Regional Event" or "Regional Burn", even those taking place in their geographic area.

The confluence of environmental, societal, and regulatory constraints at Burning Man creates a unique set of community standards that are radically different than those most participants have previously engaged with in their mostly overdeveloped, neoliberal communities. Nafstad et al.'s work on community psychology contends that neoliberalism, or late-stage capitalism, reframes community as a marketplace of individuals.

> People [in these communities] are thus primarily conceptualized as entirely self-interested, competitive and independent individuals, in the end driven by asocial greed, while aspects of common goods and collective arrangements are left unconsidered and ultimately banished from the various social systems which constitute the individual's ecological environment.
> (Nafstad et al., 2009)

By contrast, the Burning Man event takes place in a remote location with harsh weather conditions, and without easy escape from discomfort, creating an absolute need for close collaboration. "The environmental constraints of the Black Rock Desert and the dependency of the organization upon volunteer contributions demand a higher level of member activity than that needed by most communities and events" (Chen, 2003). In the Black Rock Desert of Nevada where the annual event occurs, there is no foliage, no natural shade from the 100+F/38+C dry heat in an unpredictable and hostile environment. The *playa* surface is the ancient dry lakebed of Lake Lahontan, one of the flattest surfaces on earth. The area is prone to high winds. Fine particulate dust storms regularly produce whiteout conditions with limited visibility. Dust storms are not only a threat to movement as art cars (mobile art installations) or bicyclists are required to remain stationary until they pass to avoid impact with unseen objects or humans, but also the small size of the dust is hazardous to respiratory health. Limited medical support is available on-site, and that consists mostly of first aid and relief from dehydration and heat exhaustion. Other medical care is available via airlift or by driving 100 miles to the nearest hospital in Reno. Every year, for various reasons, people die during or after the event – from accidents, pre-existing medical conditions, and other circumstances related to participation in the event. Even the growing tourist populations that visit the event on art tours, or stay in the relatively luxurious Plug and Play hotel accommodations built for them, must contend to some degree with the demands of the physical environment. The affordances of place are pivotal at Burning Man.

Only in recent years has Internet access been available on-site. Access to power has challenged the event since the early days. Prior to the development of commercially available LEDs and EL wire, all lighting beyond a flashlight was generally fire. In the early days there were a few installations that required gasoline generators for electric power, such as the neon tubing built into The Man to illuminate its outline at night. In Figure 13.2 the thin parallel lines of lighting visible along the raised arms of The Man is neon tubing. In 2007, a solar voltaic array was installed to meet the demand for power, replacing the gasoline generator to light The Man's neon outline.

Technological, social, and local environmental contexts of the annual Burning Man event impact sustainable practice in the geographically connected communities beyond the event. For example, in 2007 the solar array from The Man was donated to the public school campus in the nearby town of Gerlach, Nevada. This collaboration led directly to the formation of Black Rock Solar, a non-profit solar energy design, installation, and renewable energy education outreach organization that today exclusively serves Northern Nevada's non-profit community. In May 2016, the most recent photovoltaic solar installation by Black Rock Solar went live at a low-income elementary school in Washoe County.

The sustainability of renewable energy generation technologies such as solar photovoltaic, solar thermal, wind, and hydroelectric is dependent on conditions within the local environment in which they are situated. Black Rock Solar does not seek to replicate their work in other environments. In its educational outreach, Black Rock Solar explicitly rejects universal generalizability in favor of on-site field trips to teach students how the system they installed was designed *for that specific location*, highlighting the relationships between the students, the school, and the installation site.

We describe the Burning Man Regional Network. We examine how the simple digital tools of email, online forums, and information sharing websites are used to support individuals and groups as they enact the Ten Principles in their communities for long-term sustainable change. The critical practices are (1) the material effects of sharing, reuse, fixing, and upcycling; (2) alternative strategies and critical reflection on waste management; and (3) heterogeneous sustainability critiques and practices extending beyond the Burning Man event.

Tools for extending the Ten Principles – humans leveraging digital networks

Volunteers use digitally mediated communication technologies as a bridge between Burning Man headquarters in the San Francisco Bay Area and burn communities around the physical and virtual world. Burning Man "Regional Contacts" are local volunteers. They are expected to be "computer literate" and "leverage technology to facilitate regular communications, including announcement and/or discussion email lists" (Burning Man Project, 2017b), in addition to possessing traditional community service skills. Most computer-mediated communication takes place through simple digital tools like email, social media, and informational webpages.

Regional Contacts support communities in more than 100 geographic locations in 40 countries across the North and South Americas, Africa, Asia, Europe, Australia, and New Zealand. Through the central Burning Man website, burners can subscribe to geographically focused mailing lists managed by Regional Contacts who are volunteers in their local community. Regional Contacts serve as gatekeepers to Burning Man Project resources including labor, digital information, hardware, and software. They are responsible for explaining and championing local cultural values, systems, and practices to others. There is an explicit focus on the principle of radical self-expression rather than replication or generalized normalization of practice to other locations, including the Burning Man event in the Black Rock Desert. "With decentralization, groups may 'duplicate' efforts. Rather than being discouraged for efficiency reasons, such concurrent projects were welcomed and even celebrated for their creativity and diversity" (Chen, 2016).

The systemic support for communal effort, civic responsibility, and self-expression provides space for engaging difference, resource sensitivity, and culturally appropriate development. Community activities differ by location and include, for example, beach clean-up days in Central and Southern California, Truckee River clean-ups in Reno, and redistributing excess burners' camping materials to the homeless in San Diego. Civic engagement may include group participation in local LGBTQ parades, voter registration campaigns, and policy change advocacy such as the challenge to the San Diego Municipal Code to allow commuting via skateboard in the city's downtown, a practice that encourages non-fossil fuel transportation. The communal effort emphasizes community celebration and support for individually inspired projects, even when the specific concerns are not seen as directly impacting others in the group. The once ubiquitous Burning Man slogan "No Spectators" referred to the ethos of local participatory culture where each person was explicitly enjoined to co-create the community experience.

Larger communities are centered in urban areas such as Los Angeles, New York, and Austin, and virtually in social media sites, as well as a presence, since 2003, in Second Life. Official regional burn communities are required to coordinate activity with the Burning Man Project, which actively protects their trademarks, copyrights, and image use including that of The Man. BURN2 is "the only virtual world Regional out of more than 100 Regional groups worldwide, and is the only Regional to burn the Man!" (BURN2, 2017). Initially created as Burning Life, with an in-world burning of a human effigy similar to The Man, no other Regional is permitted this use. "This unique virtual Regional spreads the culture and Ten Principles of Burning Man year-round in Second Life," (BURN2, 2017).

A public invitation to the APIs and ongoing negotiations in community "ownership"

Participants regularly create public projects, host social celebrations, and develop new digital applications. In the spirit of decommodified participatory culture, participants make them freely available. Burning Man Project, however, actively protects a long list of trademarks and copyrights central to burner identity. The legal constraints apply not only to for-profit corporate use but also use by not-for-profit burn communities. For example, in 2014 Burning Man Project sued, and, in 2016 won, trademark protection against the not-for-profit Canadian art collaborative Burn BC. Current popular digital communication tools include mobile applications like iBurn that began as a GitHub repository and is now available for free through the Apple and Android App Stores, and websites like BurnerMap (burnermap.com). "If you're a software developer looking to create the next hot Burning Man app, [the Get Involved page of the Burning Man website invites you to] take a look at [their] Innovation section, including APIs into historical Burning Man datasets." The 1NN0VAT10N page explains that "[m]uch like Black Rock City provides a context and container for physical creations and manifestations by artists, performer and builder collectives, [the] Innovation section facilitates the development of technical and virtual tools, built by our community for our community."

The 1NN0VAT10N page is meant to *support, not to control, replicate, or generalize* unique divergent technological developments. The page explains "that the applications and links shown below are operated by application providers over which Burning Man exercises no control. The descriptions below have been supplied by the application providers, and Burning Man neither endorses these applications nor warrants that they operate as described". In the Terms of Service for Burning Man APIs and datasets, the developer agrees to a number of terms around noncommercial use in addition to the requirement that, "[i]n keeping with Burning Man's Gifting Principle, you agree that your App will be made available to participants and other interested users **free of charge**" (emphasis in original). It is important to reiterate the active legal protections Burning Man Project has in place. The Burning Man Project regularly litigates against not only commercial ventures that seek to leverage the brand, but also local burn communities who argue for the right to claim use.

Material effects

Burn communities are concerned with material effects. Salvage, reuse, sharing for maximal use, burn event legacy (i.e., heirloom status through passing an object and its history to future generations), fixing, and upcycling are all topics of interest to the burn community. These practices are on display during all burn events in the form of interactive art installations, open large- and small-scale theme camps, and the infrastructures on which these events rely. They are often not just visible, but explicitly called out as evidence of belonging. Longevity of use is valued; materials that have been to more burns are prized. Resell hubs aimed at burners will often include the number of burn events an item has been to cited as evidence of its robustness, durability, and belonging. This narrative can lead to traditional sustainability debates around the trade-offs between increased efficiency and material reuse. Do you buy the new LED flashlight or keep using the old one until it dies?

Each year after the Burning Man event in Black Rock City, and to a smaller degree after local burn community events, there is outreach offering tangible materials to public school teachers, libraries, service organizations, and museums. In all seven communities, burners offer materials to public school teachers within their local communities. These materials include reusable items like tarps and leftover art supplies. The outreach is conducted digitally through email group lists, the official Burning Man Project digital newsletter *Jack Rabbit Speaks*, through post-event announcement lists, posts to social media groups, in-person discussions at local build spaces, schools, and museums, during meetings and through flyers and handwritten notes on physical message boards. These materials are often passed along to friends, and friends of friends. All seem to eventually find a place where they are utilized. After the Burning Man event every year for at least the past four years, one or more public school teachers requested donations of general, and sometimes very specific artwork, materials, tools, and/or in-person volunteer knowledge sharing for their classroom. These requests are explicitly encouraged by burners not directly involved in either supplying or requesting these materials, through sharing email messages, replies to posts, and in the case of social media, likes and shares.

By redirecting materials from landfills and long-term storage facilities where leftover paints dry out before they can be used or fabrics are damaged due to improper storage, the burn community not only mitigates potential environmental pollution, but also challenges material waste management practices and supports unique and evolving relationships with local public education institutions. The extent to which these practices impact communities and the local environment is dependent on the specific materials at hand, the individual people taking action, and the local context. What works in one school or classroom may not work for another. Materials are limited and sourced from individuals or project groups. The impact of the particular material donation or knowledge sharing experience is unique to that time and place though community relationships grow stronger over time and repeated interaction.

Burner build spaces support practices of material, tool, and knowledge reuse and sharing. Similar to MIT's Fab Labs, burner build spaces are networked knowledge sharing places, where people gather socially and creatively to use tools and machinery and share knowledge through formal and informal learning. However, unlike the Fab Lab model that requires a common set of tools and processes to be included as a node in the network, burner build spaces are required only to share the ethos, the Ten Principles. The resources and practices vary and evolve as the populations ebb and flow. Physical characteristics change with interest and available materials, even in build spaces within the same burn community. The Reno Generator's website informs their visitors:

> We're an inclusive art space in greater Reno, Nevada, Sparks to be exact, for anyone who wants to make art and be part of a creative community. The Generator is an oasis of decommodification. We don't buy. We don't sell. We

dream. We convene. We create. We make. We're inspired by the magic and inspiration we see in all its enormity at Burning Man each summer, and we want to keep it alive all year. We combine forces and share resources in the name of creative growth and community involvement for all. We have industrial equipment and tools, a powerful network of creative minds, 3 phase power, and 34,000 square feet of workspace. We share our collective know-how and we hope to learn from yours. We're here to encourage professional artists and beginners alike to make art and to learn from each other as a creative community.

(The Reno Generator, 2017)

The invitation explicitly invokes decommodification, one of the Ten Principles, as it defines sustainable local neighborhood community practices as community norms.

In keeping with the ethos, burner build spaces are generally not-for-profit and open to the public,

[b]ut some of the most established spaces that are associated with the art community, such as NIMBY and American Steel (now defunct) in the Bay Area, are not non-profits. They're artist owned and managed spaces, yes, and they bend over backwards to help the community, absolutely, but they are not non-profits. One could argue that they're not "burner build spaces" because they're not owned or operated by the Burning Man Project or a Regional, but . . . that would inadequately represent the esteem they are held in by the community, and the services they have provided. The particular governance model is less important than the relationship they have with the community they serve.

(Caveat, 2017)

All build spaces in the seven communities described in this chapter are not-for-profit, host upcycling workshops, maintain physical community message boards, and explicitly support material reuse from projects within the community. For example, in March 2015, the first author hosted a welding workshop in one local community space for participants to learn basic welding techniques while converting a former steel food storage barrel into a functional burn barrel.

This build space has also hosted a series of glass workshops for upcycling used beverage containers, as well as laser-cutter training and self-defense workshops. All burner community places, brick and mortar build spaces and events, and virtual places including the Burning Man website's ePlaya and Spark forums, local burn email groups, and social media group pages, feature a sub-space that attends to objects and services that are intended for reuse without financial compensation. These include objects to be diverted from landfills and artisan knowledge burners seek to share with others and vary depending on local resources.

Figure 13.3 A participant learning to adjust the torch while practicing welds on a thick piece of steel that is resting on one of the former food storage containers

Photo credit: a.m. tsaasan, 2015

Critical reflection on waste management practices

The practice of burning waste, while common in the burn community and in rural communities across the United States, is not without conflict. Disagreements during events, as well as official and unofficial discussion in online forums and social media are common in burn communities. Topics like climate change and personal health risks are set forth as arguments against burning waste, and burning art, and even burning The Man. Those opposed raise issues of carbon emissions and toxic off-gassing from low-temperature burning (compared to industrial incineration). Proponents of waste combustion cite convenience, landfill leakage, and the ethical concerns in appropriating native habitats for waste. Opportunities for debate arise at various moments of social friction, e.g., when event policy changes are communicated, when first-time attendees pose questions to online forums, and any time waste burning occurs (though debates occur more often around community burn platforms rather than individual or camp burn barrels). For example, on several occasions, the color of smoke rising from a community burn platform signaled to concerned participants an opportunity for debate on the environmental impacts and trade-offs between combustion and alternative waste management practices.

Policy changes around combustion are clearly communicated during the annual process of registering art for installation. When the current year's art installation submission process is opened, conversations with collaborators and community members consider the pros and cons of new and old policies. Changes to burning art policy over the years included the introduction of mandatory mitigation measures to protect the playa surface. Past practices of burning directly on the ground lead to "burn scars". Burn scars are hard-packed sediment patches, formed by radiant heat coupled with smoke and debris that remain for years until dug up and hauled away. Participants in the remediation process often become the most vocal proponents of new mitigation policies. Mitigation practices include dismantling art and placing it in communal burn platforms, and if the installation is intended to burn in place, the artist is required to either build a platform on which to place their work or use decomposed granite for heavy, large-scale installations like that of The Man and the annual iteration of the architectural memorial installation The Temple.

The Temple burn tradition was instigated upon the death of a crew member of the 2000 installation, Temple of the Mind. "Not long before the event, one of the crew members died in a motorcycle accident. The [Temple] structure was dedicated as a memorial to him, and also as a space for anyone to leave memorials and inscriptions which would be burned with the structure . . ." (The Temple Crew, 2017). The filigree-like panels were hauntingly beautiful during the day and formed striking negative spaces when the Temple burned. The Temple burn was transformative for many, including the first author, and has inspired lead artist David Best, the crew, and the community to bring a collaboratively constructed Temple to burn on-site every year since. Though Temples have varied in outline, footprint, material use, and project lead, they are always designed as public places to honor mourning, and they always burn the night after The Man. A broad range of objects with personal significance are placed within the Temple. From a beloved grandparent and former pastor's bible, to a healing former partner's wedding dress, the now thousands of objects placed on or within the Temple each year are sacred, and thus, appear to be exempt from combustion debates.

Open burning is limited to approved art installations, a few public burn platforms, and within camps in raised burn containers. Campfire-style burning directly on the ground is prohibited. Hauling out the ashes and leftover debris is an explicit expectation. In some cases, such as that of The Man and Temple burns, ashes and debris are collected by participants for personal use or to share with others.

Heterogeneous community impacts spread beyond Burning Man

When burners return to their default worlds, many bring a commitment to engage their local community in some way that reflects the Ten Principles. As Katherine Chen described in 2003, "Members attempt to apply practices to the larger [local] society, ranging from voluntary pickup of others' trash to formally organizing local art events that reclaim cityscapes for non-commercial enjoyment." After the main burn event there is an upsurge in donations of time to local service organizations and schools to share expertise and new skills acquired during an event. Some burners seek out or found environmental organizations, public education institutions, or arts-based service organizations. Others direct time and energy to civic change and public service. In San Diego, for example, each July a free interactive public arts event with a focus on material reuse is held in collaboration with local artists and community activists who continue improvements to Chicano Park. The event website is the primary interface to the public. Here general information and frequently asked questions are posted along with a free form text message box asking if "there [is] anything specific you have in mind that you'd like to be part of . . ." (FIGMENT, 2017). Adverse to radio buttons and predefined choices in digital interactions, these communicative forms favor the flexibility of responses afforded by open fields, where rhetorical style, method, and level of commitment can be holistically evaluated by local project leaders and community organizers.

The Burning Man website and electronic newsletter *Jack Rabbit Speaks* include curated content primarily contributed by paid and unpaid staff. The image gallery and the discussion forums are the few areas where individuals can add content without prior human moderation. A majority of the participatory community's communication material is individual content. It was through the online *Burning Man Journal* in 2007 that Tom Price promoted the practice of annual environmental cleanup on May 5. Tom wrote "Burning Man Regional groups and Burners Without Borders from all over have already organized over 20 cleanup events – so you can either join an existing group . . . or get your own together." The article is still up today and contains the addresses for planned clean-up sites, inline links to local event online web pages, and personal phone numbers and email addresses of event leaders. Individuals in voluntary leadership roles entrust their personal contact information to the broader network. Individuals seeking to participate have direct access. This person-to-person network is not scalable in the sense that by using the digital tools the human nodes can increase capacity, but it is transferable in that more individuals are explicitly invited to add themselves and their specific local environmental clean-up project to the list.

Burners without Borders (BWB) emerged from the Regional Network, became an international non-profit organization, and is now part of the Burning Man

Project umbrella non-profit, "[s]upporting volunteers from around the world in innovative disaster relief solutions & community resiliency projects" (BWB Mission, 2017). In 2005 Hurricane Katrina hit during the Burning Man event, and news of the destruction and the need to mitigate environmental damage spread rapidly through Black Rock City. A group of burners, many of whom had worked on The Temple build crew at Burning Man, traveled to Biloxi, Mississippi to aid in the reconstruction of a Vietnamese temple destroyed in the hurricane. Burners used the Regional Network's flexibility to meet local demands that were not being met by government or large-scale NGOs. "While the rest of the world's attention was focused on New Orleans, the small group decided to head towards Biloxi, which had been hit just as hard but was receiving little help" (BWB Mission, 2017). They initially named themselves the Temple to Temple crew. After completing work on the temple in Biloxi, they moved to Pearlington and renamed themselves Burners Without Borders (BWB). "BWB was the only volunteer group on the Gulf Coast to receive a donation of heavy machinery, which enabled them to put Pearlington three years ahead of the relief effort in their region" (BWB History, 2017). Financial, material, and human resources were donated from local burn communities around the planet. BWB currently provides small-scale, $100–1000USD grant funding every April to projects lead by burners in their local communities. "BWB is a meme that is propagating throughout the world with every project initiated by individuals like you. With the support of the BWB infrastructure including the website, list-serves, and coaching on community organizing, you can easily initiate civic projects in your local community" (Burners Without Borders, 2017). What began as a specific local action to mitigate environmental damage became an international network supporting local change globally.

Conclusion

Many participants who travel to the remote, resource-scarce Black Rock City become sensitized to local environmental constraints that necessitate a high degree of self-reliance and communal effort to not only survive, but also to thrive. Some participants developed more sustainable geographically appropriate practices. Black Rock Solar is one such example where the need to power an art installation off-grid became a photovoltaic education, advocacy, and installation non-profit serving Northern Nevada.

Within the Burning Man Regional Network, simple digital tools are used to support change in local practices responsive to existing community resources. Local systems emphasize and orient support around community relationships. Critical burn community practices around material reuse include tool and knowledge sharing, fixing and upcycling, and alternative strategies for and critical reflection on waste management. Burn community practices are punctuated by annual patterns of participation in burn events. The cycles provide a framework for regularly reflecting, reconsidering and restructuring existing relationships and available resources within local geographically bounded areas. For example, the relationships between local burn communities and public school teachers, which persist over years, yet often consist of only one-off contributions of materials and time that take place after a burn event.

The computer literacy sufficient to act as a Regional Contact in the Burning Man Regional Network is relatively low. However a high degree of fluency in the Ten Principles is absolutely necessary for participation. In these creative communities, *simpler digital tools are leveraged by a network of communities*. Volunteers serve as gatekeepers to Burning Man Project resources including labor, digital information, hardware, and software, and are responsible for explaining and championing local cultural values, systems and practices to others. The participatory network is used to enact the Ten Principles in culturally appropriate sustainable change. We see sustainable practices highlighted throughout the network that suit the geographically situated tangible and intangible resource streams of a specific location. While human-computer interaction research is often focused on the viability of new product development, this chapter has reflected on social changes supported by older digital communication tools put into the service of human communities.

References

BURN2. (2017). *About*. Retrieved April 20, 2017, from http://burn2.org/about

Burners Without Borders. (2017). *Create a Project*. Retrieved April 20, 2017, from www.burnerswithoutborders.org/create-a-project

Burning Man Project. (2017a). *First-timer's guide*. Retrieved April 19, 2017, from https://burningman.org/event/preparation/first-timers-guide/

Burning Man Project. (2017b) *Regional Leadership Criteria*. Retrieved September 6, 2017, from http://regionals.burningman.org/becoming-a-regional-contact/regional-leadership-criteria/.

Chen, K. (2003). Coordinating contributing members: How the Burning Man organization forms an "alternative" artistic community in the Nevada Black Rock Desert. In J. W. Robinson, K. A. Harder, H. L. Pick, & V. Singh (Eds.), *People shaping places shaping people Environmental Design Research Association (EDRA) Proceedings 34* (pp. 56–61).

Chen, K. K. (2016). "Plan your burn, burn your plan": How decentralization, storytelling, and communification can support participatory practices. *The Sociological Quarterly*, 57(1), 71–97.

FIGMENT. (2017). *Volunteer*. Retrieved April 20, 2017, from http://sandiego.figmentproject.org/volunteer

The Generator. (2017). *Welcome to The Generator!* Retrieved April 19, 2017, from www.therenogenerator.com/

Magister, C. (2017) Research and review correspondence June 2017.

Nafstad, H. E., Blakar, R. M., Carlquist, E., Phelps, J. M., & Rand-Hendriksen, K. (2009). Globalization, neo-liberalism and community psychology. *American Journal of Community Psychology*, 43(1–2), 162–175.

Tavakoli, H., Khashayar, P., Amoli, H. A., Esfandiari, K., Ashegh, H., Rezaii, J., & Salimi, J. (2011). Firework-related injuries in Tehran's Persian Wednesday eve festival (Chaharshanbe Soori). *The Journal of Emergency Medicine*, 40(3), 340–345.

The Temple Build Crew. (2017). *How we got started*. Retrieved April 19, 2017, from www.thetemplecrew.org/about.html.

Response 4 Sustainability futures and the future of sustainable HCI

Yolande Strengers

Sustainability, Bendor reminds us, is all about the future. So too is design. Somewhat surprisingly then, the field of sustainable HCI has adopted a narrow vision of the future, skirting around the "Big Questions" (Beck and Stolterman, 2017), and maintaining a "remarkable heterogeneity of methods, orientations, and approaches" (DiSalvo et al., 2010: 1975). Recurring and idealized sustainability characters or users, such as "Resource Man" (Strengers, 2013), sit alongside a common suite of 'desirable' and 'helpful' technologies (such as home automation, micro grids, and ecofeedback devices) expected to reorient societies towards a relatively predictable and known sustainable future.

While not intending to dismiss or undermine these efforts, one effect of this focus has been minimal opportunity for ordinary people to imagine and co-create their own futures. The authors of the three chapters in this section turn this around with their emphasis on "everyday designers" (Desjardins et al.), "worldmaking activities" (Bendor), and "older digital technologies" (Tsaasan & Nardi) in realizing different visions and versions of sustainability. Together they constitute a refreshing shake-up of HCI's assumptions and expectations about the future, and the role designers can, do, and should play in bringing this about.

While not all chapters in this section explicitly engage with 'the future', I reflect here on how they can be read in this light, by identifying three crosscutting themes that open up a field of possibilities for imagining and co-creating sustainable futures.

The continually unfolding future

In their analysis of the role of everyday designers as placemakers in the ongoing design and creation of a campervan project and community garden, Desjardins and her colleagues draw our attention to the constant unfinishedness involved in making place. These authors challenge the idea that sustainability is about reaching or achieving a desired state – a common focus of sustainability research, policy, and programs (e.g., achieving a greenhouse gas target or resource reduction benchmark). Instead, they draw our attention to the continual processes of futuring, as everyday designers improvise, adapt, and co-create sustainable places.

This constant unfolding is also evident in Tsaasan and Nardi's analysis of Burning Man events in the United States of America and the communities that have formed around them. These authors emphasize an ongoing dynamism and flux involved in the making of sustainable futures, where locality, materials, and interpretations of the Burning Man's Ten Principles coalesce to support ongoing change and improvement towards sustainability. Similarly, Bendor, who engages explicitly with notions of futuring through his concept of worldmaking interactions, emphasizes a process of continual discovery, in which the public is encouraged to "collectively reformulate a sense of what is possible and hopefully rediscover its capacity to make, unmake and remake the world".

The never-ending-ness of sustainability evident here also draws these authors' attention to longevity in design – a key aspect of sustainable futures emphasized by both Desjardins et al. and Tsaasan and Nardi. For Desjardins and colleagues, longevity is manifest through the continual negotiation involved in retrofitting and updating places such as camper vans and community gardens, in which unused or "unuseful" materials are adapted, recycled or reappropriated in ongoing lifecycles. Similarly, in Tsaasan and Naardi's analysis of Burning Man communities, longevity is evident in the continual unfolding and dynamism involved in the events and communities themselves, and in the longevity of material use and the value they accumulate through ongoing community practices and repeat appearances at "burns".

Multiple futures

If we accept that a sustainable (or any) future is constantly unfolding over long periods of time, we must also take seriously Bendor's contention that there are multiple or plural futures. It is this understanding that inspires Bendor to develop a form of interaction which proliferates "diverse, thought-provoking, inspiring futures". This results not only in a multiplicity of futures, as Bendor argues, but also allows the capacity of the public to enact and imagine alternatives. Plurality then, is a way of potentially performing multiple realities. We see this enacted in the case studies presented by Desjardins et al. and Tsaasan and Nardi, whereby local varieties of placemaking or Burning Man events and communities constitute different possible versions of sustainability at the same time, and over time. The futures are unfolding.

However, none of the sustainability scenarios discussed emerge from a blank slate. Bendor and his colleagues design a playfully interactive space for worldmaking interactions that includes some familiar utopian and dystopian ideas. Tsaasan and Nardi describe how Burning Man communities fiercely protect their brand and the Ten Principles which underpin the event's ethos (even though the interpretation of these is continually changing), and Desjardins et al. describe forms of placemaking that are familiar practices in many parts of the world (e.g., community gardens and retrofitting vans). Thus, these authors reveal how societies and communities are oriented towards sustainable futures that are both familiar and divergent.

Creative and imaginative futures

A final key theme across these chapters is their emphasis on creativity and imagination in moving towards sustainable futures. None start with a static definition of sustainability, nor do they have a desired behavior or end goal they seek to see realised. They focus on the ways in which everyday designers, citizens or publics imagine and create their own futures, drawing attention to the ways interaction designers might support this ongoing process. Tsaasan and Nardi explore the role of 'older' or 'simpler' technologies in encouraging the circulation of ideas, recruiting people to "sustainable geographically appropriate practices", and reproducing shared understandings and competencies regarding sharing, fixing, and upcycling. Such technologies allow locally situated practices to remain locally coordinated, as well as enabling the sharing and circulation of practices to the broader Burning Man Network.

Creativity and imagination are also key aspects of the case studies described by Desjardins et al., whereby communities or individuals are given the space and freedom to co-design their own places through continual improvisation and adaptation. Bendor embraces the concept of creativity in a different way, seeking to design creative spaces that allow for open-ended reflection and playful interactions that leave opportunity for imagining multiple possible futures. As all the authors note, this is not necessarily easy, nor is it considered comfortable and desirable by all who partake in the process. For example, Bendor comments that some participants in his worldmaking experiment left the experience feeling "utterly confused, frustrated and even agitated". This is not a reason to be put off though. Others are keen to engage in the "ontological multiplicity" Bendor and his colleagues invite them to imagine and co-create for alternative futures.

Peeking around the corners

We cannot know what the future holds, but we can help facilitate it. Sustainable HCI designers are in an ideal position to agitate for its ongoing, multiple, and creative realization. As the authors of these chapters suggest, this involves dropping pre-conceived and generalized understandings of sustainability as a fixed goal or state, valuing differences and adaptation in local practice, and inspiring reflection and collaboration towards alternative futures. This opens up the possibility for sustainability to be something which we have not yet imagined, emerging from situations, places, events, and practices where ordinary people are invited to the party.

References

Beck, J., & Stolterman, E. (2017). Reviewing the Big Questions literature; or, should HCI have big questions? In *Proceedings of Designing Interactive Systems*. New York: ACM.

DiSalvo, C., Sengers, P., & Brynjarsdóttir, H. (2010). Mapping the landscape of sustainable HCI. In *Proceedings of the SIGCHI Conference on Human Factors in Computing Systems* (pp. 1975–1984). New York: ACM.

Strengers, Y. (2013) *Smart energy technologies in everyday life: Smart Utopia?* London: Palgrave Macmillan.

Photo Essay 6

Locked gate (2017). A locked gate serves as a metaphor for the ongoing search for a key to unlocking the principles by which sustainability can be integrated into interaction design research and practice. It is not obvious how to enter this gate. Can the "*No Entry*" sign that discourages entry as a matter of policy be ignored? Can we climb over the arched ironwork at the top designed to prevent climbing? Can we reach through the two door handle–level holes to unlock the unseen lock? Reflection on Remy: *Communicating SHCI Research to Practitioners and Stakeholders* and editors Hazas and Nathan's comments about the ongoing actions our community can take to advance the ideas in this book.

Eli Blevis

Epilogue

Mike Hazas and Lisa P. Nathan

The contributions in this collection provide generative ways of thinking through myriad questions, approaches, and perspectives related to the design and use of digital technology, human thriving, and the survival of other life forms with whom we share Earth. As editors, we are first to acknowledge that this text is not a comprehensive account of theories of sustainability, key sustainability thinkers, sustainable human-computer interaction (SHCI) scholarship to date, or areas that still need to be addressed. Then there are the many researchers we hoped could contribute, but had other commitments. Finally, despite the hundreds of references throughout this book, there is relevant thinking that gets little treatment. Examples include: how much needs to be done[1] to avoid the release of CO_2; food and water systems; acknowledging the connections between ecological deterioration and social injustice; reckoning with population stabilization or decline;[2] and more direct confrontation of collapse[3] scenarios. As one member of our advisory board put it, "If the book were to represent the field, we would still have work to do. But, one book cannot be everything."

In the introduction, we mentioned a thread of despair, perhaps despondency that a reader may recognize across a number of the chapters in this collection. Those working in the area of SHCI are relatively few, yet this volume brings forward the significant differences in how this loose affiliation of researchers approach their work. Despite these differences, we (the editors) suspect that the thread is tied to a shared frustration with the 'business-as-usual' approach to conducting research in human-computer interaction (HCI). In HCI, there is a tendency to fetishize the blistering edge, feeding the expectation that researchers should constantly realign their work to trending topics to attract funding, top students, and favorable reviews of their innovative scholarship. This prevailing attitude is often at odds with and significantly undercuts sustainability research.

Ten years into it, there are grumblings that there isn't much to show. The hubris of the field reveals itself: one of the striking things about HCI is its can-do, change-the-world overnight attitude. But we would point out that for generations, dedicated people within and outside academia have been working to minimize humanity's influence on our environment. Why would HCI scholars expect themselves to have an immediate, discernable impact on globally intertwined phenomena? For complicated problems that take decades, if not longer, to address, researchers need to engage deeply over the span of their careers (not weeks or months) in the same area, developing expertise and connections to others working on related issues. Much needs to be done to change the expectations and support for SHCI research moving forward.

Drawing on thinking from other fields is one of the most engaging strengths of SHCI; and the broader perspective that often results is important for tackling the problem. However, we would point out that continually drawing on new fields, however exciting or insightful, for a single research effort is at odds with the depth of knowledge and allowance for reflection that a more focused set of theories, methods, and analysis can facilitate. We would also encourage SHCI researchers to continue to look carefully at how to transfer their outcomes to areas of real-world impact. Even if that transfer ultimately fails, we will slowly get better at negotiating that interface to impact, and it will become easier to reject the expectation that all research in HCI should be able to change the world within the span of the annual conference review cycle.

Notes

1 Saul Griffith, 2009. http://longnow.org/seminars/02009/jan/16/climate-change-recalculated/
2 https://earthswords.wordpress.com/2012/02/27/the-impact-of-negative-population-growth/
3 See for example Tim Bennett and Sally Erickson's 2007 documentary film, *What a Way to Go: Life at the End of Empire*.

Index

Note: Italicized page numbers indicate a figure on the corresponding page. Page numbers in bold indicate a table on the corresponding page.

ACM Code of Ethics and Professional Conduct 11
activism-based organization 81
actor-network theory 183
aggregated choices 211
Airstream traveling home 185
amateur gardeners 91
anthropogenic greenhouse gas emissions 47
artificial intelligence (AI) 107
Association of Computing Machinery SIGCHI Conference on Human Factors in Computing Systems (CHI) 5, 9; *see also* communicating SHCI research; human-computer interaction (HCI); sustainable human-computer interaction (SHCI)
augmentation of human-product relations 23
autobiographical design project 193

Bamako Convention (1991) 143
bartering 95
Basel Convention (1989) 143
behavior change support systems 167
Binzen, William 218
biodiversity 141
black box design 22
Blevis, Eli 5
Blum, Andrew 93
bottom-up analysis 76
boundary-negotiating objects 39–40
Brundtland Commission 205
Burn BC 223
Burners without Borders (BWB) 228–229

Burning Man: community impact beyond 228–229; digital technology 221–223; introduction to 217–221, *218*, *219*; material effects 223–226, *226*; summary of 229–230; waste management practices 226–228
Burning Man Journal 228
Burning Man Regional Network 217, 219, 230

capitalism 19, 49, 160
carbon dioxide 31, 93, 227
Carsen, Rachel 205
Chaharshanbe Suri ceremony 218
China supply chain 80
civic engagement at Burning Man 222
climate change 31, 156
'closing' effect 36
cloud metaphor 93
cognitivism 32
collapse computing 96
collapse informatics field 51, 88, 91
communicating SHCI research: framing of 134–137; introduction to 129–130; opportunities and limitations 136–137; to stakeholders 131–133; summary of 137; target audience engagement 131–134; theory-practice gap 129, 130–131, 136; translating knowledge 134–136
community-centered approach to urban planning 190
community garden case study 194–195, 196–201, *197*, *199*, *201*
companion planting 113

238 *Index*

computational agroecology 90–94
computing education for sustainability 47–48
computing essentials 93–94
computing-supported efficiencies 48
computing within LIMITS 88
concept-driven direction 24
Conference of the Parties (COP) 143
consistent futures 208–209
consumption understanding 176
continually unfolding futures 231–232
continuous adaptation 72
cornucopian paradigm 162
corporate purpose in political economy 96–98
creative futures 233
crisis informatics field 88

Danish military case study 173–175
defuturing 213
de-growth movement 75, 81
design as futurescaping 206–208
design fictions 40
Design Futuring (Fry) 81
Designing Publics framework 192–193
despair, in SHCI research 6, 235
diagnostic knowledge of repairs 21
digital artifacts 154
digital technology: approaches relying on 182; design and use of 235; engineering and 116; environmental impacts of 51; HCI link 53; introduction to 3–4, 22; leveraging networks 221–223; sustainability in design of 61; use of 45
Dispossessed, The (Le Guin) 208
dissatisfaction, in SHCI research 6
Dodge v. Ford (1919) 97
Dogwood Initiative 209
do-it-yourself (DIY) project 193–194
domain-dependent characterization of sustainability 111
domain experts 170
Dourish, Paul 34

earned functionality 23
East Africa Compliant Recycling facility 143
eco-efficiency 54
eco-feedback technology 33, 133
EcoHouse 172–173, 175, 183
economic development/social justice–first development 72
economic exchange value of devices 21

economic growth 87, 160, 162
economic sustainability 107
ecovillages 90
education: computing education for sustainability 47–48; political economy 94–96; Software Engineering for Sustainability 118
e-Learning solutions 174
elimination design strategies 20
emotional barriers 179
enabling effects 106
Encyclical on Climate Change and Inequality (Pope Francis) 206
energy demand reduction 141
engagement in human-product relations 23
engineering mindset *vs.* limits to growth 117–118
environmental awareness 18
environmental collapse 95
environmental footprint 63
environmental public policy (EPP): environmental policy and HCI 142–147; EPEAT-certified products 146; green public procurement 140–141, 145–147; introduction to 140–141; opportunities and challenges 147–148; public policy and HCI 141–142; summary of 148; waste electrical and electronic equipment 140–145
environmental sustainability 17, 50–51, 75, 107
environment-first development 72
EPEAT-certified products 146
equality in sustainable interaction design 22–25
ethics and sustainability 46
ethnographic observations 23
European Environment Agency 55
everyday design 192
e-waste 86, 132, 143, 179
extended producer responsibility (EPR) 143

facts as ideology 160–161
Fairphone (FP) project: challenges to 122–123; defined 76–77; introduction to 25, 75–76; invention and disposal 78–79; objectives, assumptions and expectation 77–78; ownership and identity 80–81; quality and equality 79–80; renewal and reuse 79
Fairtrade certification scheme 160
"fairware" principles 76–77

fast-industrializing nations 72
Firefly Staircase case study 171–172
food sovereignty 92
fossil fuels 31
Framework for Strategic Sustainable Development 51
Francis, Pope 206
freecycling 95
Free Software communities 22
free trade 160
Fry, Tony 81
functional requirements (FR) 108
future-facing technologies/practices 95

geek heresy 48
geoengineering 161
GitHub repository 223
Global Affairs Canada 141
Goodman, Nelson 206
Google 55
greenhouse gas emissions 47, 91, 93
green public procurement (GPP) 140–141, 145–147, 183
Grudin, Jonathan 87

happiness, in SHCI research 98
HCI4D field 88, 145
heirloom systems 25
high-performance computing (HPC) 107
high-tech design/management 92
histories of human-product relations 23
human-computer interaction (HCI): environmental policy and 142–147; participatory design and governance 182–184; political economy and 88–89, 123; public policy and 141–142; response to 179–181; software engineering for sustainability 103; summary of 235–236; sustainability concept 107; *see also* sustainable human-computer interaction
human-machine interactions 7
human-nature dichotomy 73
human-product relations 23

ICTD field 88
ignorance-based worldview 55–56
imaginations in SHCI 65–66
imaginative futures 233
immediate effects 106
improvisations in SHCI 64
improvised performance of practice 39
indigenous knowledge 91

'individual behavior change' philosophy 61
individual sustainability 106
industrial agriculture 91
industrialism 74
information and communications technologies (ICT) 182
information processing 32
inteconnections in SHCI 64–65
interaction design research 117
Interaction paradigm 31–32
interactive technologies 17, 193
Interface Carpet story 52
intergenerational equity 50
intergovernmental organizations 141
International Monetary Fund 92
Issues of Governance Studies 96

Jacobs, Jane 189, 190

knowledge-based worldview 55–56
KTH Royal Institute of Technology, Sweden 154

Le Guin, Ursula 208
LGBTQ rights 96
liberalism 160
lifecycle assessment (LCA) 106
Limits to Growth, The report 74, 205
linking intervention and disposal 17–20
longevity and placemaking 195–197, *196*, *197*
longevity considerations 23–24
long-term sustainability 87

March for Science 7
Merchants of Doubt (Oreskes, Conway) 92–93
mobilization of practice approaches 5–6
model-reality relations 209
modular devices 130
monoculture 91
Moore's law 19
multiple futures 232
multiplicity in placemaking 199–201, *200*, *201*

Nagata, Kai 209
Nairobi Declaration 143
natural modes and reflection 14
nature/humanity rhetoric 82
neoliberalism 220
Nest thermostat 130

non-functional requirements (NFR) 108–109
non-profit organizations 141

obsolescence: introduction to 17–18, 20; placemaking and 192; planned obsolescence 86–87
online communities 3
ontological diversity 38
Open Source Software communities 22
Organisation of African Unity 143

participatory design 48, 169–170, 182–184, 192–193
perceived durability of human-product relations 23
perceived worth 23
persuasive technology (PT): case studies 170–175; discussion over 175–176; introduction to 166–167; participatory design and 169–170; placemaking and 191; response to 180; summary of 176; for sustainability 167–169
placemaking: community garden case study 194–195, 196–201, *197*, *199*, *201*; discussion about 201–202; everyday design 192; introduction to 189–190; longevity and 195–197, *196*, *197*; multiplicity in 199–201, *200*, *201*; participatory and social design 192–193; summary of 202; sustainable interaction design 191–192; sustainable placemaking 190–191; themes of 195–201; unfinishedness 197–199, *198*, *199*; van conversion case study 193–200, *196*, *198*, *200*
planetary boundaries 154
planetary capacity to sustain life 72
planned obsolescence 86–87
plant guild composer 113
political ecology 72, 73
political economy: action needed for 88–98; computational agroecology 90–94; corporate purpose 96–98; education and 94–96; human-computer interaction 88–89, 123; introduction to 86–88; summary of 98
polybrominated biphenyls (PBB) 144
polybrominated diphenyl ethers (PBDE) 144
positive socio-economic transformation 63
PowerHouse 167, 182
practice-orientation 31–32, 34–35

premature obsolescence 17–18
Price, Tom 228
public policy and HCI 141–142
public-private partnerships 143

quality in sustainable interaction design 22–25
Quigley, Ocean 208

renewal and reuse 20–22
requirements engineering for sustainability (RE4S) 107, *109*, 109–110
resilience thinkers 73
Resilient Smart Gardens with Permaculture and Sensors project 111, 113–116, *114*, *115*, 122
reuse as is, category 21
rooftop garden 67
Rubric of Material Effects (RoME) 17–18, 20

satisfaction, in SHCI research 132
Scandinavian workplace democracy movement 169
scientifically defensible 208–209
scientific progress 162
scientist-activists 162
Scott, James C. 182
second-order effects 93
selfies 1
Shove, Elizabeth 37
Silent Spring (Carsen) 205
SimCity game 208
small farmers 91
smartphone industry 76–78
social design 192–193
social media/networking 3
social practice theory 31, 34
social sustainability 106
socio-ecological transformation: computing and 46–53; introduction to 44–46, *45*; models and metrics 55–57; overview of 53–54; summary of *57*, 57–58
socio-natural arrangements 73
sociopolitical analysis 86
Software Engineering for Sustainability (SE4S): background and foundations 105–110, *109*; concepts of 105–107; defined 104–105; education implications 118; engineering mindset *vs.* limits to growth 117–118; future exploration 117; human-computer

interaction 103; introduction to 103–105; requirements engineering for sustainability 107, *109*, 109–110; requirements of 108–109; Resilient Smart Gardens with Permaculture and Sensors project 111, 113–116, *114*, *115*; response to 122–124; scope and relating impact 114–116, *115*; stakeholder model *110*, 110–114, **112–113**, *114*; summary of 116–118; sustainability and 104–105

sorrow, in SHCI research 156

SourceMap 130, 131

Stakeholder Reference List 111, **112–113**

stakeholders: communicating SHCI research 131–133; HCI interconnections 64–65; social practice theory 35; software engineering *110*, 110–114, **112–113**, *114*

start-up organization 81

'state-of-the-art' IoT infrastructure 48

Steampunk culture 38, 40, 90, 211

Stout, Lynn A. 96–97

structural effects 106

subject-matter experts 170

substitution effects 48

sufficiency and longevity 23

supply chain to China 80

survivalists 90

sustainability: concepts of 105–107; development of 72; Fairphone (FP) project 25, 75–81; within human-computer interaction 74–75; as innovation 52; introduction of 71–72; methodology of 75–76; moving carousel of 72–74; requirements with 108–109; single understanding of 176; summary of 81–82; technical sustainability 107; *see also* teaching sustainability at technical universities

sustainability futures: design as futurescaping 206–208; plurality of 205–206; prediction *vs.* invention 212–213; response to 231–233; scientifically defensible and consistent futures 208–209; worldmaking interactions 206, 209–212

Sustainability in an Imaginary World installation 210, 212

Sustainable Development Goals 57

sustainable human-computer interaction (SHCI): alternative futures and design 9–11; beyond HCI audiences 7–9; broader scope for design 33–35; contradictions in 3–4; deliberation and reflection 4–5; imaginations 65–66; improvisations 64; interconnections 64–65; introduction to 31–33; limits of 6–7; reflections on 5–6; response to 180; within society 63–66; summary of 11, 41, 235–236; temporality and change 35–41; user as active producer 189; *see also* communicating SHCI research; human-computer interaction

sustainable interaction design (SID): from individual to system 61–62; introduction to 17–18; linking intervention and disposal 17–20; origin of 74, 76; placemaking 191–192; quality and equality 22–25; renewal and reuse 20–22; revisiting principles of 78–81; summary of 25

system boundaries 62

target audience engagement 131–134

teaching sustainability at technical universities: colleague difficulties 157–158; facts as ideology 160–161; friction with 155–161; hyperloop 161–162; introduction to 154–155; loss, fear and sadness 155–157; meeting student expectations 159–160; rethinking future careers 158–159; reverberations 162–163

technical sustainability 107

technological solutions 162

temporality and change 35–41

theory-practice gap 129, 130–131, 136

top-down initiatives (cockpitism) 73

toxic material dumping 142–143

transdisciplinary perspective 61

Transition Town movement 52, 90

true biological ecosystems 91

Tube: A Journey to the Center of the Internet, The (Blum) 93

UbiGreen 167, 168, 182

un-designed movement 48, 75

unfinishedness and placemaking 197–199, *198*, *199*

United Nations' Intergovernmental Panel on Climate Change (IPCC) 205

urban mining 77–78

urban planning 190

user-centered design 86

US National Parks Service 56

values-based commitment 54
van conversion case study 193–200, *196, 198, 200*
virtual discussion groups 4
Virtues of Ignorance, The (Vitek, Jackson) 55–56

waste electrical and electronic equipment (WEEE) 140–145, 180, 183
waste management practices 226–228
WaterBot 167, 168, 182

Watson, Matt 34
Whyte, William 189, 190
Wishbone Design Studio 52
work conditions 78
Workshop on Computing within Limits 6–7
World Bank 92
worldmaking interactions 206, 209–212
World Trade Organization 92
WWOOFer (voluntary workers on organic farms) 158